D1051921

A BRIEF TOUR OF HUMAN CONSCIOUSNESS

From Impostor Poodles To Purple Numbers

V. S. Ramachandran, M. D.

PI PRESS
NEW YORK

Pi Press

An Imprint of Pearson Education, Inc.
1185 Avenue of the Americas, New York, New York 10036

Published by Pearson Education Inc.

Pi Press offers discounts for bulk purchases. For information contact U.S. Corporate and Government Sales, 1-800-382-3419, or corpsales@pearsontechgroup.com. For sales outside the U.S., please contact International Sales at international@pearsoned.com.

Printed in the United States of America

10 9 8 7 6 5 4

The BBC in association with Profile Books Ltd.
First published in Great Britain in 2003 by
Profile Books Ltd.
58a Hatton Garden
London Ec1N 8lX

Library of Congress Cataloging-in-Publication Data

Ramachandran, V. S.
A brief tour of human consciousness: from impostor poodles to purple numbers / V. S. Ramachandran.
p. cm.
Includes bibliographical references and index.
ISBN 0-13-187278-8 (pbk. : alk. paper) 1. Consciousness. 2. Cognitive neuroscience.
3. Neurobehavioral disorders. I. Title.
QP411.R32 2005
612.8'2—dc22

2005014952

ISBN 0-13-187278-8

Pearson Education Ltd.
Pearson Education Australia Pty., Limited
Pearson Education Singapore, Pte. Ltd.
Pearson Education North Asia Ltd.
Pearson Education Canada, Ltd.
Pearson Educación de Mexico, S.A. de C.V.
Pearson Education — Japan
Pearson Education Malaysia, Pte. Ltd.

For my parents, Vilayanur Subramanian and Vilayanur Meenakshi

For Diane, Mani and Jaya

For Semmangudi Sreenivasa Iyer

For President Abdul Kalam, for launching the youth of our country into the new millennium

For Shiva Dakshinamurthy, Lord of Gnosis, music, knowledge and wisdom

Contents

Preface

My goal in writing this book has been to make neuroscience—the study of the brain—more accessible to a broad audience, to "workingmen," as Thomas Huxley would have said. The overall strategy is to investigate neurological dysfunction caused by a change in a small part of a patient's brain and ask: Why does this patient display these curious symptoms? What do the symptoms tell us about the workings of the normal brain? Can a careful study of these patients help us explain how the activity of a hundred billion nerve cells in the brain gives rise to all the richness of our conscious experience? I have chosen to focus both on areas in which I have worked directly (such as phantom limbs, synesthesia and visual processing) and ones that have a broad interdisciplinary appeal, in order, ultimately, to bridge the gap that now separates C. P. Snow's "two cultures"—the sciences and the humanities.

The book emerged from the annual BBC Reith lectures that I delivered in Great Britain in 2003. It was an honor for me to be invited to give these lectures, the first physician/experimental psychologist to do so since they were begun by Bertrand Russell in 1948. In the last five decades these lectures have enjoyed a distinguished place in the intellectual and cultural life of the Western world. I was delighted to accept the invitation, knowing that

I would be joining a long list of previous lecturers whose works inspired me as a teenager: Peter Medawar, Arnold Toynbee, J. Robert Oppenheimer, John Galbraith and Russell, to mention only a few. I realized that theirs would be a tough act to follow, given their towering stature and the pivotal role that many of them played in defining the intellectual ethos of our age. Even more daunting was the requirement that I would have to make the lectures not only interesting to the specialist but also intelligible to the "common people," thereby fulfilling Lord Reith's original mission for the BBC. Given the enormous amount of research on the brain, the best I could do was to provide an impressionistic survey rather than try to be comprehensive. In doing this I was worried I might have oversimplified many of the issues involved and so run the risk of annoying some of my specialist colleagues. But as Lord Reith himself once said, "There are some people whom it is one's duty to annoy!"

Chapter 3 (based on the third lecture) deals with an especially controversial subject, the neurology of artistic experience, "neuro-aesthetics," that is usually considered off limits by scientists. I take a stab at it just for fun and to indicate how a neuroscientist might approach this problem. I make no apology for the fact that it is speculative. As Peter Medawar said, "all good science begins as an imaginative excursion into what might be true." Speculation is fine, provided it leads to testable predictions and so long as the author makes it clear when he is merely speculating—skating on thin ice—as opposed to when he's on solid ground. I have taken pains to preserve this distinction throughout the book, often adding qualifying remarks in what have become extensive endnotes.

There is also a tension in the field of neurology between the "single case study" approach, the intensive study of just one or two patients with a syndrome, and sifting through a large number of patients and doing a statistical analysis. The criticism is sometimes made that it is easy to be misled by single strange cases, but this is nonsense. Most of the syndromes in neurology that have stood the test of time—for example, the major aphasias (language disturbances), amnesia (explored by Brenda Milner, Elizabeth Warrington, Larry Squire and Larry Weiskrantz), cortical color blindness, neglect, blindsight, "split brain" syndrome (commissurotomy), etc.—were initially discovered by a careful study of single cases, and I don't know of even one that was discovered by averaging results from a large sample. The best strategy, in fact, is to begin by studying individual cases and then to make sure that the observations are reliably repeatable in other patients. This is true for a majority of findings described here, such as phantom limbs, the Capgras (impostor) delusion, synesthesia and neglect. The findings are remarkably consistent across patients and have been confirmed in several laboratories.

Many colleagues and students often ask me when I became interested in the brain and why. It is not easy to trace the lineage of one's interests, but I'll give it a shot. I have been interested in science from about the age of eleven. I remember being a somewhat lonely child and socially awkward—although I did have one very good science playmate in Bangkok: Somthau ("Cookie") Sucharitkul. I always felt companioned by Nature and perhaps science was my "retreat" from the social world with all its arbitrariness and mind-numbing conventions. I spent a lot of time collecting seashells and geological specimens and fossils. I enjoyed dabbling in ancient archaeology, cryptography (the Indus script), comparative anatomy and palaeontology; I found it endlessly fascinating that the tiny bones inside our ears, which we mammals use for amplifying sounds, had originally evolved from the jawbones of reptiles. As a schoolboy I was passionate about chemistry and often mixed chemicals just to see what would happen (a burning piece of magnesium ribbon could be plunged into water—it would continue to burn underwater by extracting the oxygen from H2O). Another passion was biology. I once tried placing various sugars, fatty acids and individual amino acids inside the "mouths" of Venus flytraps to see what triggered them to stay shut and secrete digestive enzymes. And I did an experiment to see if ants would hoard and consume saccharin—showing the same fondness for it as they do for sugar. Would the saccharin molecule "fool" their taste buds the same way it fools ours?

All these pursuits, Victorian in inspiration, are quite remote from the neurology and psychophysics I specialize in now. Yet those childhood preoccupations must have left a mark that profoundly influences my adult personality and my style of doing science. While engaged in such pursuits, I felt that I was in my own private playground, my own parallel universe inhabited by Darwin and Cuvier and Huxley and Owen and William Jones and Champollion. For me these people were more alive than most "real" people I knew. Perhaps this escape into my own private world made me feel special rather than isolated, "weird" or different. It allowed me to rise above the tedium and monotony—the humdrum existence that most people call a normal life—to a place where, to quote Russell, "one at least of our nobler impulses can escape from the dreary exile of the actual world".

Such an escape is especially encouraged at the University of California, San Diego, an institution that is both venerable and vibrantly modern. Its neuroscience program was recently ranked number one in the country by the National Academy of Sciences. If you include the Salk Institute and Gerry Edelman's Neurosciences Institute, there is a higher concentration of eminent neuroscientists in La Jolla's "neuron valley" than anywhere else in the world. I can't think of a more stimulating environment for anyone interested in the brain.

Science is most fun when it is still in its infancy, when its practitioners are still driven by curiosity and it hasn't become "professionalized" into just another nine-to-five job. Unfortunately this is no longer true for many of the most successful areas of science, such as particle physics or molecular biology. It is now commonplace to see a paper in *Science* or *Nature* cowritten by thirty authors. For me this takes some of the joy out of it (and I imagine it does for the authors too). This is one of two reasons I instinctively gravitate toward old-fashioned Geschwindian neurology, where it is still possible to ask naive questions starting from first principles—the kinds of very simple questions that a schoolboy might ask but are embarrassingly hard for experts to answer. It's a field where it's still possible to do Faraday-style research and come up with surprising answers. Indeed, many of my colleagues and I see it as an opportunity to revive the golden age of neurology, the age of Charcot, Hughling Jackson, Henry Head, Luria and Goldstein.

The second reason I chose neurology is more obvious; it's the same reason you picked up this book. As human beings we are more curious about ourselves than about anything else, and this is a research enterprise that takes you right into the heart of the problem of who we are. I got hooked on neurology after examining my very first patient in medical school. He was a man with a pseudo-bulbar palsy (a kind of stroke), who alternately laughed and wept uncontrollably every few seconds. It struck me as an instant replay of the human condition. Were these just mirthless joy and crocodile tears, I wondered? Or was he actually feeling alternately happy and sad, the same way a manic-depressive might, but on a compressed timescale?

We will be considering many such questions throughout the book: What causes phantom limbs? How do we construct a body image? Are there artistic universals? What is a metaphor? Why do some people see musical notes as colored? What is hysteria? Some of these questions I will answer, but the answers to the others remain tantalizingly elusive, such as the big question: "What is consciousness?" But whether I answer them or not, if this book at least whets your appetite to learn more about this fascinating field, it will have served its purpose. Endnotes, a glossary, a bibliography, and an index are provided at the end for the benefit of those who wish to probe these topics more deeply. As my colleague Oliver Sacks said of one of his books: "The real book is in the endnotes, Rama."

I would like to dedicate this work to the patients who volunteered to endure hours of testing at our center. I have often learned more from my conversations with them, despite their damaged brains, than from my learned colleagues.

I

A Pain in the Brain

The history of mankind in the last three hundred years has been punctuated by major upheavals in human thought that we call scientific revolutions — upheavals that have profoundly affected the way in which we view ourselves and our place in the cosmos. First there was the Copernican revolution — the notion that far from being the center of the universe, our planet is a mere speck of dust revolving around the sun. Then there was the Darwinian revolution, culminating in the view that we are not angels but merely hairless apes, as Thomas Henry Huxley once pointed out. And, third, there was Freud's discovery of the "unconscious" — the idea that even though we claim to be in charge of our destinies, most of our behavior is governed by a cauldron of motives and emotions of which we are barely conscious. Your conscious life, in short, is nothing but an elaborate post-hoc rationalization of things you really do for other reasons.

But now we are poised for the greatest revolution of all — understanding the human brain. This will surely be a turning point in the history of the human species for, unlike those earlier revolutions in science, this one is not about the outside world, not about cosmology or biology or physics, but about ourselves, about the very organ that made those earlier revolutions possible. And I want to emphasise that these insights into the human brain will have a profound impact not just on scientists but also on the humanities, and indeed they may even help us bridge what C. P. Snow called the two cultures — science on the one hand and arts, philosophy and humanities on the other. Given the enormous amount of research on the brain, all I can do here is to provide a very impressionistic survey rather than try to be comprehensive. The five chapters cover a very wide spectrum of topics, but two recurring themes run through all of them. The first broad theme is that by studying neurological syndromes which have been largely ignored as curiosities or mere anomalies we can sometimes acquire novel insights into the functions of the normal brain — how the normal brain works. The second theme is that many of the functions of the brain are best understood from an evolutionary vantage point.

The human brain, it has been said, is the most complexly organized structure in the universe and to appreciate this you just have to look at some numbers. The brain is made up of one hundred billion nerve cells or "neurons" which form the basic structural and functional units of the nervous system (Figure 1.1). Each neuron makes something like one thousand to ten thousand contacts with other neurons and these points of contact are called synapses. It is here that exchange of information occurs.

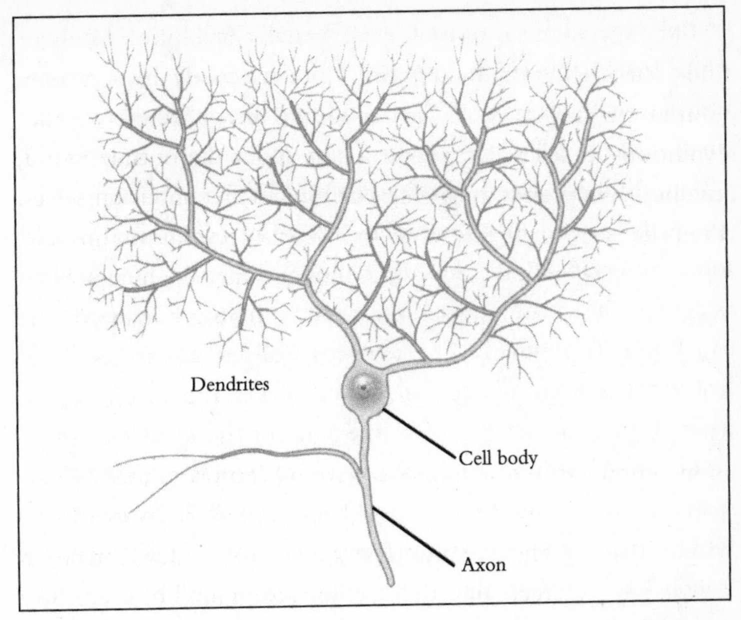

Figure 1.1 *Drawing of a neuron showing dendrites which receive information from other neurons and a single long axon that sends information out to other neurons.*

Based on this information, it has been calculated that the number of possible permutations and combinations of brain activity, in other words the numbers of brain states, exceeds the number of elementary particles in the known universe. Even though it is common knowledge, it never ceases to amaze me that all the richness of our mental life—all our feelings, our emotions, our thoughts, our ambitions, our love lives, our religious sentiments and even what each of us regards as his or her own intimate private self—is simply the activity of these little specks of jelly in our heads, in our brains. There is nothing else. Given this staggering complexity, where does one even begin? Well, let's start with some basic anatomy. In the twenty-first century most people have a rough idea of what the brain looks like. It has two mirror-image halves, called the cerebral hemispheres, and resembles a walnut sitting on top of a stalk, called the brain stem. Each hemisphere is divided into four lobes: the frontal lobe, the parietal lobe, the occipital lobe and the temporal lobe (Figure 1.2). The occipital lobe in the back is concerned with vision. Damage to it can result in blindness. The temporal lobe is concerned with hearing, emotions and certain aspects of visual perception. The parietal lobes of the brain—at the sides of the head—are concerned with creating a three-dimensional representation of the spatial layout of the external world, and also of your own body within that three-dimensional representation. And lastly the frontal lobes, perhaps the most mysterious of all. They are concerned with some very enigmatic aspects of the human mind and human behavior such as your moral sense, your wisdom, your ambition and other activities of the mind which we know very little about.

There are several ways of studying the brain, but my approach is to look at people who have had some sort of damage or change to a small part of their brain. Interestingly, people who have had a small lesion in a specific part of the brain do not suffer an across-the-board reduction in all their cognitive capacities; no blunting of their mind. Instead there is often a highly selective loss of one specific function while other functions are preserved intact, which is a good indication that the affected part of the brain is somehow involved in mediating the impaired function. I could cite many examples, but here are some of my favorites.

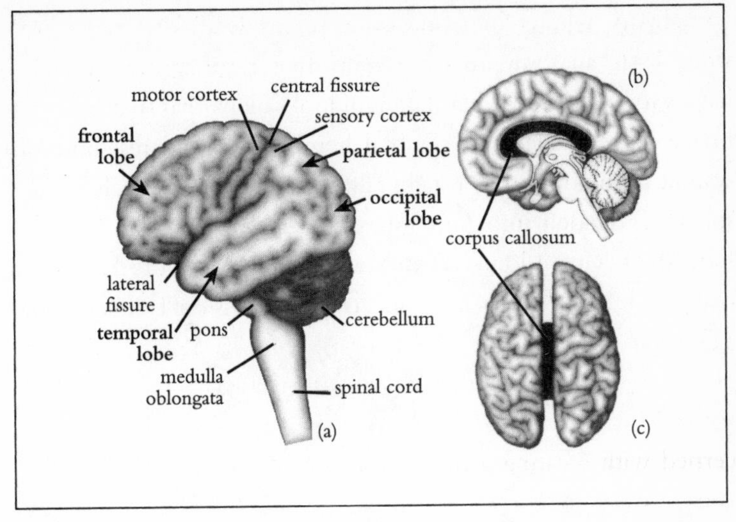

Figure 1.2 *Gross anatomy of the human brain. (a) Shows the left side of the left hemisphere. Notice the four lobes: frontal, parietal, temporal and occipital. The frontal is separated from the parietal by the central or rolandic sulcus (furrow or fissure), and the temporal from the parietal by the lateral or sylvian fissure. (b) Shows the inner surface of the left hemisphere. Notice the conspicuous corpus callosum (black) and the thalamus (white) in the middle. The corpus callosum bridges the two hemispheres. (c) Shows the two hemispheres of the brain viewed down the top. (a) Ramachandran; (b) and (c) redrawn from Zeki, 1993.*

First, prosopognosia, or face blindness. When a structure called the fusiform gyrus in the temporal lobes is damaged on both sides of the brain, the patient can no longer recognize people's faces (Figure 1.3). The patient can still read a book, so is not blind, and is not psychotic or mentally disturbed in any way but is simply no longer able to recognize people by just looking at the face.

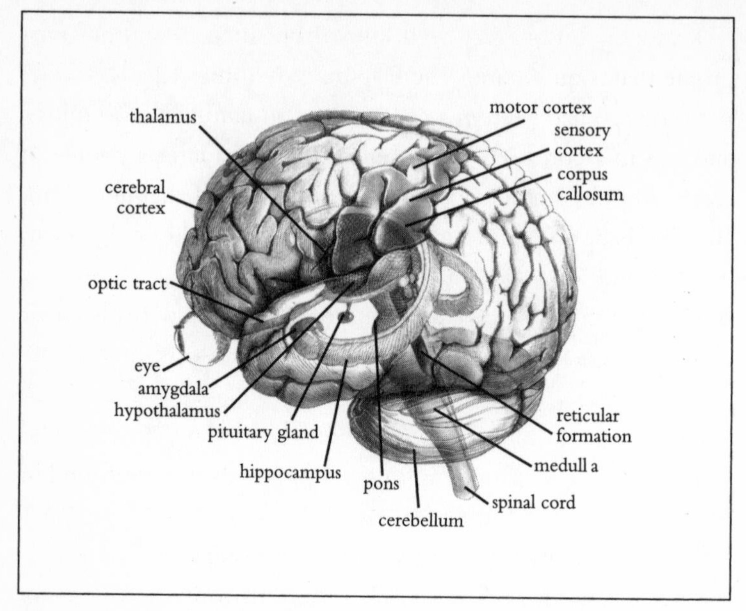

Figure 1.3 *Artist's rendering of a brain with the outer convoluted cortex rendered partially transparent to allow inner structures to be seen. The thalamus (dark) can be seen in the middle, and interposed between it and the cortex are clusters of cells called the basal ganglia (not shown). Embedded in the front part of the temporal lobe you can also see the hippocampus (concerned with memory). In addition to the amygdala, other parts of the limbic system such as the hypothalamus can be seen. The limbic structures mediate emotional arousal. The hemispheres are attached to the spinal cord by the brain stem (consisting of medulla, pons and midbrain), and below the occipital lobes is the cerebellum, concerned mainly with coordination of movements and timing. The fusiform gyrus—concerned with processing faces—is in the inner side of the temporal lobe in the bottom. The amygdala—which receives signals from the fusiform—can be seen clearly in the diagram. From* Brain, Mind and Behaviour *by Bloom and Laserson (1988) by Educational Broadcasting Corporation. Used with permission from W. H. Freeman and Company.*

Prosopognosia is very well known but there is another syndrome that is quite rare—the Capgras syndrome. A patient I saw not long ago had been in a car accident, sustaining a head injury, and was in a coma. He came out of the coma after a couple of weeks and was quite intact neurologically when I examined him. But he had one profound delusion—he would look at his mother and say, "Doctor, this woman looks exactly like my mother but she isn't, she is an impostor." Why would this happen? Bear in mind that this patient, who I will call David, is completely intact in other respects. He is intelligent, alert, fluent in conversation and not emotionally disturbed in any other way.

The standard explanation for the Capgras delusion—found in older psychiatry textbooks—is a Freudian one. According to this view, during infancy all of us men have a strong sexual attraction toward our mothers—the so-called Oedipus complex. But as we grow up the cortex becomes more highly developed and inhibits or "represses" these latent urges originating in the limbic emotional core of the brain. Thank God for that, otherwise we would all be permanently sexually aroused by our mothers! But then, along comes a blow to the head and these repressed sexual urges come flaming to the surface, so that suddenly the patient finds himself sexually aroused by his mother. The only way he can rationalize these forbidden feelings is to say, "If it's mom . . . why am I sexually turned on? It must be an impostor." Hence the delusion. This is an ingenious explanation—as indeed all Freudian explanations are—but it doesn't work. I have seen a patient having the same delusion not only about his mother but also about his pet poodle! "This isn't Fifi, doctor . . . it's some other dog that looks like Fifi." Clearly the Freudian explanation cannot explain this without invoking something absurd like "the latent bestiality of all humans," which would be too far-fetched even by the notoriously lax intellectual standards of Freudian psychology.

So, what really causes Capgras syndrome? To understand this disorder, you have to first realize that vision is not a simple process. When you open your eyes in the morning, it's all out there in front of you and so it's easy to assume that vision is effortless and instantaneous. But in fact within each eyeball, all you have is a tiny distorted upside-down image of the world. This excites the photoreceptors in the retina and the messages then go through the optic nerve to the back of your brain, where they are analyzed in thirty different visual areas. Only after that do you begin to finally identify what you're looking at. Is it your mother? Is it a snake? Is it a pig? And that process of identification takes place partly in a small brain region called the fusiform gyrus—the region which is damaged in patients with face blindness or prosopognosia. Finally, once the image is recognized, the message is relayed to a structure called the amygdala, sometimes called the gateway to the limbic system, the emotional core of your brain, which allows you to gauge the emotional significance of what you are looking at. Is this a predator? Is it prey which I can chase? Is it a potential mate? Or is it my departmental chairman I have to worry about, a stranger who is not important to me, or something utterly trivial like a piece of driftwood? What is it?

In David's case, perhaps the fusiform gyrus and all the visual areas are completely normal, so his brain tells him that the woman he sees looks like his mother. But, to put it crudely, the "wire" that goes from the visual centers to the amygdala, to the emotional centers, is cut by the accident. So he looks at his mother and thinks, "She looks just like my mother, but if it's my mother why don't I feel anything toward her? No, this can't possibly be my mother, it's some stranger pretending to be my mother." This is the only interpretation that makes sense to David's brain, given the peculiar disconnection.

How can an outlandish idea like this be tested? My student Bill Hirstein and I in La Jolla, and Haydn Ellis and Andrew Young in England, did some very simple experiments measuring galvanic skin response (see chapter 5).[1] We found—sure enough—that in David's brain there was a disconnection between vision and emotion as predicted by our theory. Even more amazing is that when David's mother phones him he instantly recognizes her from her voice. There is no delusion. Yet if an hour later his mother were to walk into the room he would tell her that she looked just like his mother but was an impostor. The reason for this anomaly is that a separate pathway leads from the auditory cortex in the superior temporal gyrus to the amygdala, and that pathway perhaps was not cut by the accident. So auditory cognizance remains intact while visual cognizance has disappeared. This is a lovely example of the sort of thing we do: of cognitive neuroscience in action; of how you can take a bizarre, seemingly incomprehensible neurological syndrome—a patient claiming that his mother is an impostor—and then come up with a simple explanation in terms of the known neural pathways in the brain.

Our emotional response to visual images is obviously vital to our survival, but the existence of connections between visual brain centers and the limbic system or emotional core of the brain also raises another interesting question: What is art? How does the brain respond to beauty? Given that these connections are between vision and emotion, and that art involves an aesthetic emotional response to visual images, surely these connections must be involved, and this forms the subject of a later chapter.

Are these intricate connections in the brain laid down by the genome in the fetus, or are they acquired in early infancy as we begin to interact with the world? This is the so-called nature/nurture debate, and is central to my next example: phantom limbs. Most people know what is meant by a phantom limb. A patient has an arm amputated because it has a malignant tumor or has been irreparably damaged in an accident but continues to feel the presence of the amputated arm. A famous example concerns Lord Nelson, who vividly felt a phantom arm long after his real one had been lost in battle. (He actually used it in a somewhat flawed argument for the existence of a non-corporeal soul. For if an arm can survive physical annihilation, he asked, why not the whole body?)

I once had a patient whose arm had been amputated above the left elbow. He sat in my office blindfolded while I gently touched different areas of his body and asked him to say where I was touching him. All went as expected until I touched his left cheek, at which point he exclaimed, "Oh my God, you're touching my left thumb," his missing phantom thumb, in other words. He seemed as surprised as I was. Touching his upper lip produced sensation in his phantom index finger, and touching his lower jaw provoked sensations in his phantom little finger. There was a complete, systematic map of the missing phantom hand draped on his face (Figure 1.4).

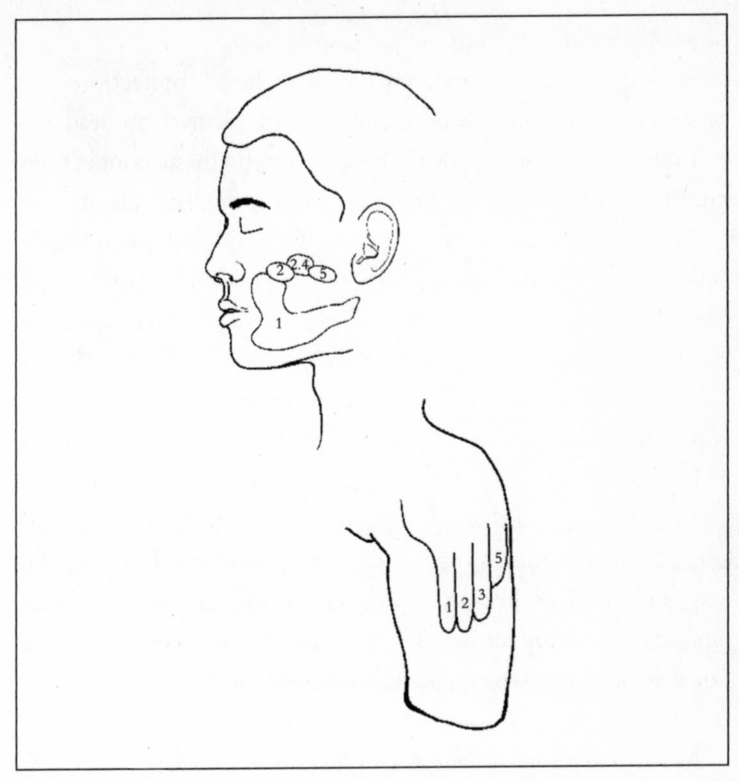

Figure 1.4 *Points on the body surface that yielded referred sensations in the phantom hand (this patient's left arm had been amputated ten years prior to our testing him). Notice that there is a complete map of all the fingers (labelled 1 to 5) on the face and a second map on the upper arm. The sensory input from these two patches of skin is now apparently activating the hand territory of the brain (either in the thalamus or in the cortex). So when these points are touched, the sensations are felt to arise from the missing hand as well.*

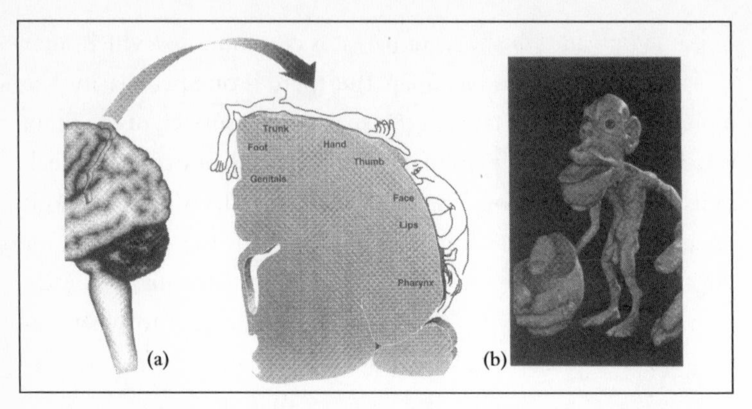

Figure 1.5 *(a) The representation of the body surface on the surface of the human brain behind the central sulcus. The homunculus ("little man") is upside down for the most part and his feet are tucked onto the medial surface (inner surface) of the parietal lobe near the very top, whereas the face is down near the bottom of the outer surface. Notice, also, that the face area is below the hand area instead of being where it should—near the neck—and that the genitals are represented below the foot. (b) A whimsical three-dimensional model of the Penfield homunculus—the little man in the brain.*

Why does this happen? Like the Capgras delusion, phantom limbs are a mystery that would have intrigued Sherlock Holmes. What on earth is going on? The answer lies, again, in the anatomy of the brain. The touch signals from the entire skin surface on the left side of the body are mapped on to the right cerebral hemisphere on a vertical strip of cortical tissue called the post-central gyrus. Actually, there are several maps but for ease of representation we can assume there is only one map, called SI, on the post-central gyrus. This is a faithful representation of the entire body surface—almost as if there were a little person draped on the surface of the brain (Figure 1.5). We call this the Penfield homunculus, and for the most part it is continuous—which, after all, is what one means by a map. But there is one peculiarity: the representation of the face on this map on the surface of the brain is right next to the representation of the hand, not near the neck in its expected position. The head is dislocated. (Why this is so is unclear; perhaps it has something to do with the phylogeny or the way in which the brain develops in early fetal life or in early infancy, but dislocated it is.) This gave me the clue to what was happening. When an arm is amputated, no signals are received by the part of the brain's cortex corresponding to the hand. It becomes hungry for sensory input and the sensory input from the facial skin now invades the adjacent vacated territory corresponding to the missing hand. Signals from the face are then misinterpreted by higher centers in the brain as arising from the missing hand.[2] The specificity of these signals is so compelling that an ice cube or warm water applied to the face will produce a cold or warm phantom digit. One patient, Victor, when the water began to trickle down his face, also felt it trickling down his phantom arm. When he raised his arm he was amazed to feel the trickle going up his phantom, contrary to the laws of physics.

To test our "remapping" or "cross-wiring" hypothesis directly we used the brain imaging technique called MEG or magnetoencephalography. This shows which parts of the brain are stimulated when various parts of the body are touched. Sure enough, we found that in Victor (and other arm amputees like him), touching his face activated not only the face area in the brain but also the hand region of the Penfield map (Figure 1.6). This is very different from what is seen in a normal brain, where touching the face activates only the facial region of the cortex.

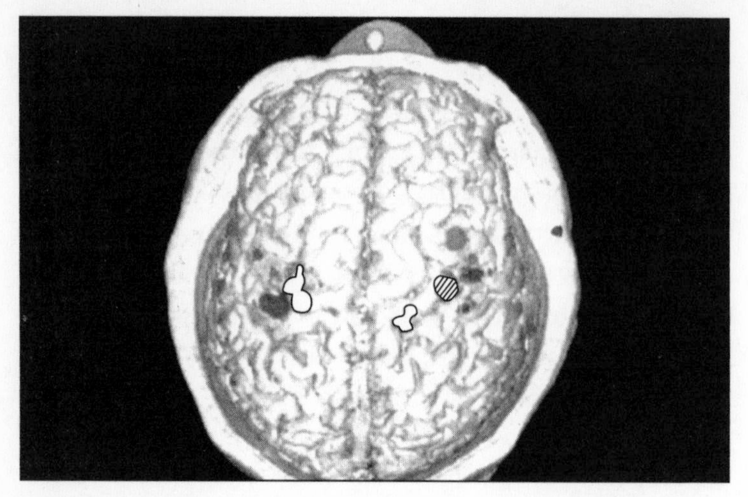

Figure 1.6 *Magnetoencephalography (MEG) image superimposed on a magnetic resonance (MR) image of the brain in a patient whose right arm was amputated below the elbow. The brain is viewed from the top. The right hemisphere shows normal activation of the right hand (hatched), face (black) and upper arm (white) areas of the cortex corresponding to the Penfield map. In the left hemisphere there is no activation corresponding to the missing right hand, but the activity from the face and upper arm has now spread to this area.*

There has obviously been some cross-wiring in Victor's brain, and this is important because it permits us to correlate the changes in brain anatomy, changes in brains' sensory maps, with the phenomenology. This link between physiology and psychology is one of the major goals of cognitive neuroscience.[3]

The discovery also has broader implications. One of the things all medical students learn is that connections in the brain are laid down in the fetus or in early infancy, and that once they are laid down, there is nothing much that can be done to change these connections in an adult. That's why when there's damage to the nervous system, such as is caused by a stroke, there is so little recovery of function. It is also why neurological ailments are notoriously difficult to treat ... or at least that's what we were taught. What I have seen flatly contradicts this view and suggests that there is a tremendous amount of plasticity or malleability even in the adult brain, and this can be demonstrated in a five-minute experiment on a patient with a phantom limb.

It isn't yet clear how this "plasticity" of body maps can be harnessed in the clinic, but I'll mention another example to show how some of these ideas can be clinically useful. Some patients can "move" their phantom limbs, and will say, "It's waving goodbye," or, "It's shaking hands with you."[4] But in many other patients the phantom arm feels "paralyzed," "frozen stiff," "in cement," or "won't budge an inch." Often the phantom hand goes into painful involuntary clenching spasms or is fixed in an awkward painful position which the patient is unable to change. We have discovered that some of these patients had pre-existing nerve damage before the amputation, for example the arm had been paralyzed and lying in a sling. After amputation the patient is stuck with a paralyzed phantom ... as if the paralysis is "carried over" into the phantom. Perhaps when the arm was intact but paralyzed, every time the front of the brain sent a command to the arm saying "move," it was getting visual feedback saying "no, it won't move." Somehow this feedback becomes imprinted on the circuitry in the parietal lobe or somewhere else in the brain. (We call this "learned paralysis.") How could this highly speculative idea be tested? Perhaps if a patient were given visual feedback that the phantom was obeying the brain's commands

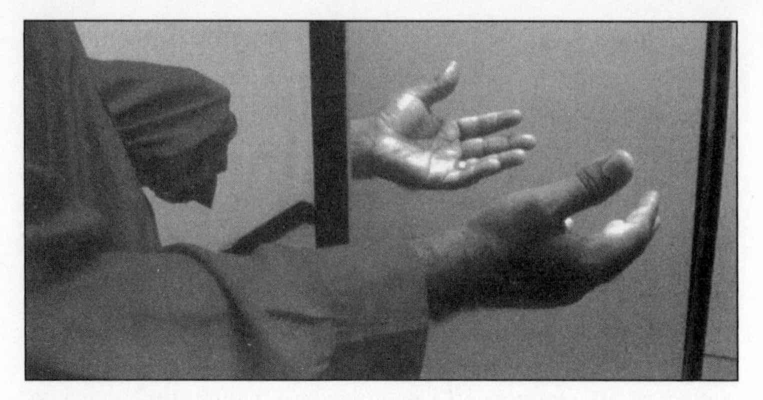

Figure 1.7 *Illustrates the "mirror box" arrangement used for resurrection of the phantom.*

the learned paralysis could be "unlearned." We propped up a mirror vertically on a table in front of a prone patient, so that it was at right angles to his chest, and asked him to position his paralyzed phantom left arm on the left of the mirror and mimic its posture with his right hand, which was on the right side of the mirror. We then asked him to look into the right-hand side of the mirror so that he saw the mirror reflection of his intact hand optically superimposed on the felt location of the phantom (Figure 1.7). We then asked him to try to make symmetrical movements of both hands, such as clapping or conducting an orchestra, while looking in the mirror. Imagine his amazement and ours when suddenly he not only saw the phantom move but felt it move as well. I have repeated this experiment with several patients, and it seems that the visual feedback animates the phantom so that it begins to move as never before, often for the first time in years. Many patients have found that this sudden sense of voluntary control and movement in the phantom produces relief from the spasm or awkward posture that was causing much of the agonizing pain in the phantom.[5]

Relief from a phantom pain using a mirror is surprising enough, but can the same trick be applied to real pain in an intact arm or leg? Even though we usually think of pain as one thing, there are at least two different types which may have evolved for different functions. Acute pain evolved to allow reflexive withdrawal from, for example, fire, and probably also to teach avoidance of harmful, pain-producing objects such as thorns. Chronic pain—as in a fracture or a gangrene—is a different thing altogether: it evolved to reflexively immobilize the arm, so letting it rest and remain out of harm's way until fully healed. Ordinarily, pain is a very useful adaptive mechanism—a gift, not a curse. But sometimes the mechanism backfires. We often see patients with a condition called "chronic regional pain type 1," which includes the bizarre clinical syndrome of "reflex sympathetic dystrophy" or RSD. In RSD patients, what begins as a minor injury—a bruise or insect sting or fracture of a fingertip—leads to the entire arm becoming excruciatingly painful, completely immobilized, inflamed and swollen—grossly out of proportion to the inciting event. And it lasts forever.

The evolutionary framework helps us to understand how this might come about. Remember, the original purpose of chronic pain is a temporary immobilization to allow recovery, so when the brain sends a motor command to the arm there is intense pain preventing further movement. This is ordinarily adaptive, but I suggest that it sometimes malfunctions and leads to what I call "learned pain": the very act of attempting to move the arm—the motor command signal itself—becomes pathologically associated with excruciating pain. As a result, even after the inciting event has long disappeared, the patient still has a pseudoparalysis caused by learned pain. In 1995 I suggested that this type of pathological chronic pain may also benefit from mirror visual feedback. Imagine the patient sees the reflection of a normal hand superimposed optically on the painfully immobilized abnormal one. If the normal hand is now moved (while partially attempting to move the painful one) the patient will see the bad arm suddenly springing to life and moving quite freely! Perhaps this would help RSD patients to "unlearn" the spurious connection in their brains between arm movement and pain—thereby eliminating the pain and returning mobility. In 1995 this was no more than a far-fetched idea but recently McCabe et al. (2003) tried the mirror procedure on nine patients in placebo-controlled clinical trials. The pain went away completely and mobility returned in many of the patients who used mirrors, whereas the control group, who used Plexiglas, experienced no benefit at all. This result is so surprising that I would have been skeptical had not Patrick Wall—arguably the world's leading expert on both pain and placebos—been one of the authors. If confirmed, this result promises a new and effective treatment for at least some patients with chronic pain.[6]

The cross-wiring in the brain that sometimes results from amputation can also occur owing to a gene mutation. Instead of the brain modules remaining segregated, they become accidentally cross-wired, resulting in a curious condition called synesthesia, first clearly documented by Francis Galton in the nineteenth century. Synesthesia, which appears to be genetically transmitted, results in a mingling of the senses. For example hearing a particular musical note might invoke a particular color: C sharp is red, F sharp is blue, etc. Visually perceived numbers can produce a similar effect: 5 might always be seen as red, 6 always green, 7 always indigo, 8 always yellow ... Synesthesia is surprisingly common, affecting about one in two hundred people. What causes this mixing of signals? A student of mine, Ed Hubbard, and I were looking at brain atlases—specifically at the fusiform gyrus, where color information is analyzed. We saw that the number area of the brain, which represents visual graphemes of numbers, is also in the fusiform gyrus, almost touching the number area. It seems likely that, just as amputation can produce cross-wiring between the face and the hand, synesthesia is caused by cross-wiring between the number and color areas in the fusiform gyrus due to an inherited genetic abnormality.

Even though synesthesia was described by Galton a hundred years ago, the phenomenon never made it into mainstream neuroscience. It was often assumed that these people were either just crazy or simply trying to draw attention to themselves. Or maybe it had something to do with childhood memory: refrigerator magnets or a learning book in which 5 was red, 6 was blue, 7 was green ... but if that were the case, how could it run in families? My colleagues and I wanted to show that synesthesia is a real sensory phenomenon, not mere imagination or memory. We devised a simple computer display, a number of black 5s scattered on a white background. Embedded among those 5s were a number of 2s forming a hidden shape (Figure 1.8). Since these are computer- generated, the 2s are just mirror images of the 5s. Most people looking at this pattern see only a random jumble of numbers, but a synesthete sees the 5s as green and the 2s as forming a red shape

Figure 1.8 *A "clinical test" for synesthesia. The display consists of 2s embedded in a matrix of randomly placed 5s. Non-synesthetes find it very hard to discern the embedded shape (in this case a triangle). Synesthetes who see the numbers as colored can detect the triangle much more easily. (Depicted schematically in Figure 4.1.)*

conspicuously visible against a forest of green (shown schematically in Figure 4.1). The fact that synesthetes can more easily identify these shapes than normal people suggests that they are not crazy but experiencing a genuine sensory phenomenon. It also rules out memory association or some high-level cognitive phenomenon. Our group in La Jolla as well as Jeffrey Gray and Mike Morgan and others in London have conducted experiments to test the idea that there is cross-wiring in the brain. We have shown that there is activation of the fusiform gyrus in the color area when these people are shown numbers in black and white. (In normal people the color area is activated only if they are shown colored numbers.)

Phantom limb, synesthesia and Capgras' syndrome can at least be partly explained in terms of neural circuitry. But I once encountered someone with an even more bizarre syndrome called pain asymbolia.[7] To my amazement, this patient responded to a pain stimulus not with an "ouch" but with laughter. Here is the ultimate irony—a human being laughing in the face of pain. Why would anyone do this? First, we need to answer an even more basic question: why does anybody laugh? Clearly, laughter is hard-wired, it's a trait in all humans. Every society, every civilization, every culture, has some form of laughter and humor (except for Germans). But why did laughter evolve through natural selection? What biological purpose does it serve?

The common denominator of all jokes is a path of expectation that is diverted by an unexpected twist necessitating a complete reinterpretation of all the previous facts—the punch-line. Obviously a sudden twist per se is not sufficient for laughter, otherwise every great scientific discovery that generates a "paradigm shift" would be greeted with hilarity, even by those whose theory had just been disproved. (No scientist would be amused if you disproved his theory; believe me, I've tried!) Reinterpretation alone is insufficient. The new model must be inconsequential. For example, a portly gentleman walking toward his car slips on a banana peel and falls. If he breaks his head and blood spills out, obviously you are not going to laugh. You are going to rush to the telephone and call an ambulance. But if he simply wipes off the goo from his face, looks around him, and then gets up, you start laughing. The reason is, I suggest, because now you know it's inconsequential, no real harm has been done. I would argue that laughter is nature's way of signaling that "it's a false alarm." Why is this useful from an evolutionary standpoint? I suggest that the rhythmic staccato sound of laughter evolved to inform our kin who share our genes: don't waste your precious resources on this situation; it's a false alarm. Laughter is nature's OK signal.

But what does this have to do with my pain asymbolia patient? Let me explain. When we examined his brain using a CT scan we found there was damage close to the region called the insular cortex on the sides of the brain. The insular cortex receives pain signals from the viscera and from the skin. That's where the raw sensation of pain is experienced, but there are many layers to pain—it is not just a unitary thing. From the insular cortex the message goes to the amygdala, encountered earlier in the context of the Capgras syndrome, and then to the rest of the limbic system, and especially the anterior cingulate, where we respond emotionally to the pain. We experience the agony of pain and take the appropriate action. So perhaps in this patient the insular cortex was normal, so he could feel the pain, but the wire that goes from the insula to the rest of the limbic system and the anterior cingulate was cut: a disconnection similar to that seen in the Capgras patient. Such a situation would produce the two key ingredients required for laughter and humor: one part of the brain signals a potential danger but the very next instant another part—the anterior cingulate—does not receive a confirmatory signal, thereby leading to the conclusion "it's a false alarm." Hence thepatient starts laughing and giggling uncontrollably.The same sort of thing happens in tickling, which may be a sort of crude "playtime" rehearsal for adult humor. An adult approaches a child with his hands extending menacingly toward vulnerable parts of the child's body but then anti-climactically deflates the potential threat with gentle stimulation and a "Koochy, koochy, koo!"This takes the same form as mature adult humor: a potential threat followed by a deflation.

The examples we have considered so far suggest that syndromes such as phantom limb, the Capgras delusion, or pain asymbolia can help us understand the neural basis, functional logic and evolution of many otherwise mysterious aspects of our minds. This is true whether you are considering little quirks of human behavior, such as our response to tickling, or the loftiest question of all—the neural basis of the self. There is something distinctly odd about a hairless, neotenous primate that has evolved into a species that can look back over its own shoulder to ponder its origins. Odder still, the brain can not only discover how other brains work but also ask questions about itself: Who am I? What is the meaning of my existence? Why do I laugh? Why do I dream? Why do I enjoy art, music and poetry? Does my mind consist entirely of the activity of neurons in my brain? If so, what scope is there for free will? It is the peculiar recursive quality of these questions as the brain struggles to understand itself that makes neurology so fascinating.

The prospect of answering these questions in the millennium is both exhilarating and disquieting, but it's surely the greatest adventure that our species has ever embarked upon.

2

Believing Is Seeing

Our ability to perceive the world around us seems so effortless that we tend to take it for granted. But just think of what's involved. You have two tiny, upside-down distorted images inside your eyeballs but what you see is a vivid three-dimensional world out there in front of you. This transformation—as Richard Gregory has said—is nothing short of a miracle. How does it come about? What is perception?

One common fallacy is to assume there is an image inside your eyeball, the optical image, exciting photoreceptors on your retina and then that image is transmitted faithfully along a cable called the optic nerve and displayed on a screen called the visual cortex. This is obviously a logical fallacy because if you have an image displayed on a screen in the brain, then you have to have someone else in there watching that image, and that someone needs someone else in *his* head, and so on ad infinitum.

The first step we must take toward understanding perception

is to forget the idea of images in the brain and think instead of transforms or symbolic representations of objects and events in the external world. Just as little squiggles of ink called writing can symbolize or represent something they don't physically resemble, so the action of nerve cells in the brain, the patterns of firing, represent objects and events in the external world. Neuroscientists are like cryptographers trying to crack an alien code, in this case the code used by the nervous system to represent the external world.

This chapter concerns the process we call seeing—and how we become consciously aware of things around us. As in chapter 1, I'll begin with some examples of patients with strange visual defects and then explore the wider implications of these syndromes for understanding the nature of conscious experience. How does the activity of neurons—mere wisps of protoplasm—in the visual areas of the brain give rise to all the richness of conscious experience, the redness of red or blueness of blue? Or the ability to tell a burglar from a lover?

We primates are highly visual creatures. We have not just one visual area, the visual cortex, but thirty areas in the back of our brains which enable us to see the world. It's not clear why we need thirty areas and not just one. Perhaps each of these areas is specialized for a different aspect of vision. For example, one area called V4 seems to be concerned mainly with processing color information, seeing colors, whereas another area in the parietal lobe called MT, or the middle temporal area, is concerned mainly with seeing motion.

The most striking evidence for this comes from patients with tiny lesions that damage just V4, the color area, or just MT, the motion area. For example, if V4 is damaged on both sides of the brain a syndrome called cortical color blindness or achromatopsia results. Patients with cortical achromatopsia see the world in shades of grey, like a black and white film, but they have no problem reading a newspaper or recognizing people's faces or seeing direction of movement. Conversely if MT, the middle temporal area, is damaged, the patient can still read books and see colors but can't tell you which direction something is moving, or how fast.

A woman in Zurich who had this problem was terrified to cross the street because she saw the cars there not as moving but as a series of static images as though lit by a strobe light in a discotheque. She couldn't tell how fast a car was approaching even though she could read its number plate and tell you what color it was. Even pouring wine into a glass was an ordeal: she could not gauge the rate at which the wine level was rising and so the wine would always overflow. Most of us cross the road or pour a drink without even thinking about it—only when something goes wrong do we realize how extraordinarily subtle the mechanisms of vision really are and how complex a process it is.

Although the anatomy of these thirty "visual" areas in the brain seems bewildering at first, there is an overall plan of organization. The message from the eyeball on the retina goes though the optic nerve to two major visual centers in the brain. One of these, which I'll call the old system, is the evolutionary ancient pathway that includes a structure in the brain stem called the superior colliculus. The second pathway, the new pathway, goes to the visual cortex in the back of the brain (Figure 2.1). The new pathway in the cortex is doing most of what we usually think of

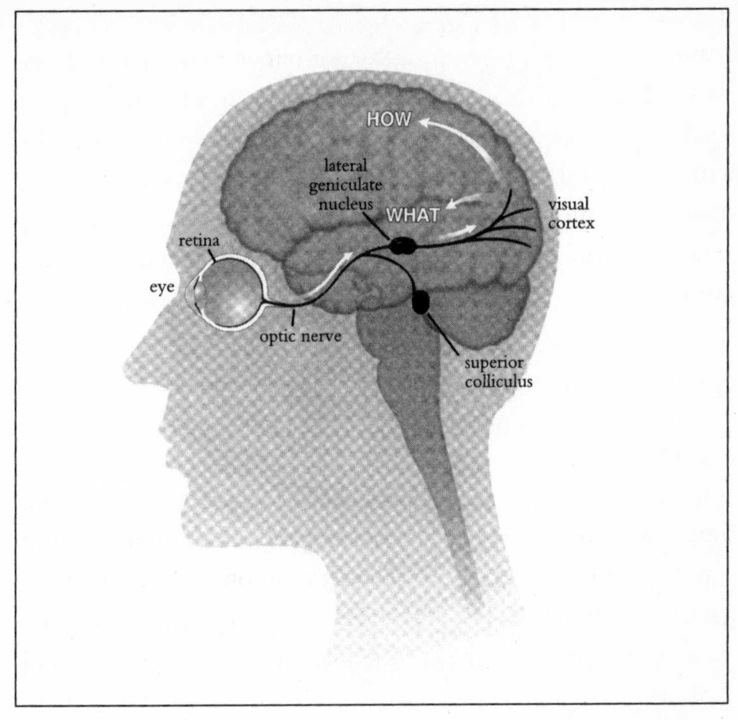

Figure 2.1 *The anatomical organization of the visual pathways. Schematic diagram of the left hemisphere viewed from the left side. The fibers from the eyeball diverge in two parallel "streams": a new pathway that goes to the lateral geniculate nucleus (shown here on the surface for clarity, though it is actually inside the thalamus, not the temporal lobe) and an old pathway that goes to the superior colliculus in the brain stem. The "new" pathway goes to the visual cortex and diverges again (after a couple of relays) into two pathways (white arrows)—a "how" pathway in the parietal lobes that is concerned with grasping, navigation and other spatial functions, and the second, "what" pathway in the temporal lobes concerned with recognizing objects. These two pathways were discovered by Leslie Ungerleider and Mortimer Mischkin of the National Institutes of Health. The two pathways are shown by white arrows.*

as vision, such as recognizing objects, consciously. The old pathway, on the other hand, is involved in locating objects spatially in the visual field, enabling you to reach out for it or swivel your eyeballs toward it. This allows the high-acuity central foveal region of the retina to be directed toward the object so that the new visual pathway can then proceed to identify the object and generate an appropriate behavior toward it: eat it, mate with it, run away from it, name it, etc.

An extraordinary neurological syndrome called blindsight was discovered by Larry Weiskrantz and Alan Cowey at Oxford and Ernst Poppel in Germany. It has been known for more than a century that damage to the visual cortex (which is part of the new visual pathway) on one side of the brain results in blindness on the opposite side. For example, a patient whose right visual cortex is damaged is completely blind to everything to the left of their nose when they are looking straight ahead (technically called the left visual field). When examining such a patient, named GY, Weiskrantz noticed something very strange. He showed the patient a little spot of light in the blind region and asked what he saw. The patient said "nothing," as would be expected. But then he asked the patient to reach out and touch the light, even though he couldn't see it.

"But I can't see it," said the patient. "How do you expect me to point to it?" Weiskrantz said to try anyway; take a guess. To the researcher's surprise, the patient reached out and pointed accurately to the dot that he could not consciously perceive. After hundreds of trials it became obvious that he could point with 99 percent accuracy, even though he claimed on each trial that he was just guessing and didn't know if he was getting it right or not. The implications of this are staggering. How can someone reach out and touch something he cannot see?

The answer is, in fact, obvious. GY has damage to his visual cortex—the new pathway—which is why he is blind. But remember, he still has the other old pathway, the other pathway going through his brain stem and superior colliculus as a backup. So even though the message from the eyes and optic nerves doesn't reach the visual cortex, given that the visual cortex is damaged, it takes the parallel route to the superior colliculus which allows him to locate the object in space. The message is then relayed to higher brain centers in the parietal lobes that guide the hand movement accurately to point to the invisible object! It's as if even though GY the person, the human being, is oblivious to what's going on, there's another unconscious being—a "zombie"—inside him who can guide his hand with uncanny accuracy.

This explanation suggests that only the new pathway is conscious—events in the old pathway, going through the colliculus and guiding the hand movement, can occur without a person being conscious of it! Why? Why should one pathway alone, or its computational style, perhaps, lead to conscious awareness, whereas neurons in a parallel part of the brain, the old pathway, can carry out even complex computations without being conscious? Why should *any* brain event be associated with conscious awareness given the "existence proof" that the old pathway through the colliculus can do its job perfectly well without being conscious? Why can't the rest of the brain do without consciousness? Why can't it all be blindsight, in other words?

We can't yet answer this question directly but as scientists the best we can do is to establish correlations and try to home in on the answer. We can make a list of all brain events that reach consciousness and a list of those brain events that don't. We can then compare the two lists and ask whether there is a common denominator in each list that distinguishes it from the other. Is it only certain styles of computation that lead to consciousness? Or perhaps certain anatomical locations that are linked to being conscious? That is a tractable empirical question which, once tackled, might get us closer to answering what the function of consciousness, if any, might be and why it evolved (just as knowing that heredity was embodied in DNA allowed us to crack the genetic code).

To compile these two lists we do need to know a great deal more about what the limits of blindsight are. How sophisticated is it? This has yet to be studied in detail. We already know from Alan Cowey and Petra Stoerig that it is capable of some degree of wavelength (color) discrimination. We know it cannot recognize faces, but can it correctly "guess" someone's expression?

One claim made for visual awareness and consciousness is that it is required for *binding* different features of an object together. If you are shown a red object moving right while a green one is simultaneously moving left, and if your color area and motion area in your brain are simultaneously signaling these attributes, how do you know which direction goes with which color? It has been suggested that consciousness is not required for the initial stage of extracting the attributes (red vs. green, left vs. right) but it *is* required to solve "the binding problem"—i.e., to know which color goes with which direction. Patients like GY can help us test this theory. The prediction would be that if he were simultaneously shown two balls on his blind side—a red one moving right and green one moving left—he should be able to tell you there is a red object and a green object and that one of them is moving right and the other left, but he won't know which is which. Or we could dispense with color and simply have two objects one below the other (both in the blind left visual field) simultaneously moving in opposite directions. Could GY tell which is which?

I should point out that the blindsight syndrome in GY seemed so bizarre that it was (and still is) greeted with skepticism by some of my colleagues. This is in part because the syndrome is very rare but also partly because it seems to violate common sense. How can you point to something you don't see? However, that is not a good reason for rejecting it because in a sense we all suffer from blindsight. Let me explain.

Imagine you are driving your car and having an animated conversation with your friend sitting next to you. Your attention is entirely on the conversation, it's what you're conscious of. But in parallel you are negotiating traffic, avoiding the pavement, avoiding pedestrians, obeying red lights and performing all these very complex elaborate computations without being really conscious of any of it unless something strange happens, such as a leopard crossing the road. So in a sense you are no different from GY: you have "blindsight" for driving and negotiating traffic. What we see in GY is simply an especially florid version of blindsight unmasked by disease, but his predicament is not fundamentally different from that of us all.[1]

Intriguingly, it is impossible to imagine the converse scenario: paying conscious attention to driving and negotiating traffic while unconsciously having a creative conversation with your friend. This may sound trivial but it is a thought experiment and it is already telling you something valuable: that computations involved in the meaningful use of language require consciousness but those involved in driving, however complicated, do not involve consciousness. It is true that sleepwalkers sometimes "talk" without (presumably) being conscious—but their mumblings are hardly like the two-way exchange of normal open-ended conversation. The link between language and consciousness is a topic we will explore further in chapter 5.

I believe this approach to consciousness will take us a long way toward answering the riddle of the benefits of consciousness and why it evolved. My own philosophical position about consciousness accords with the view proposed by the first Reith lecturer, Bertrand Russell, that there is no separate "mind stuff" and "physical stuff" in the universe: the two are one and the same. (The formal term for this is neutral monism.) Perhaps mind and matter are like the two sides of a Möbius strip that appear different but are in fact the same.

So much for the messages in the new visual pathway. Now let us turn to the other pathway, the old pathway which goes to the colliculus and which mediates blindsight and projects to the parietal lobe in the sides of the brain. The parietal lobes are concerned with creating a symbolic representation of the spatial layout of the external world. The ability that we call spatial navigation—avoiding obstructions, dodging a snowball, catching a football—all of these abilities depend crucially on the parietal lobes.

Damage to the right parietal lobe produces a fascinating

syndrome called neglect, in a sense the converse of blindsight. The patient no longer moves his eyes toward the object, which is looming toward him from the left, and he can no longer reach out and point to it or grab it. But he is not blind to events on the left side of the world because if you draw his attention to an object there he can see it perfectly clearly and can identify it. The best description of the neglect syndrome is an indifference to the left side of the world. A patient suffering from this syndrome will eat only from the right side of a plate and leave the food on the left side uneaten. Only if the patient's attention is drawn to the uneaten food will he eat what's there. A man will shave only the right side of his face; a woman will apply makeup similarly. And if you give the patient a sketchpad and ask her to draw a flower the result will be only half a flower—the right half (Figure 2.2).

Neglect is caused by damage to the right hemisphere, and the patient is also usually paralyzed on the left side because the right hemisphere of the brain controls the left side of the body. I wondered if it would be possible to "cure" neglect. Could patients be treated by making them pay attention to the left side of the world which they are ignoring?

I again hit on the idea of using a mirror, as in the case of phantom limbs in chapter 1. I had the patient sit on a chair and then I stood to her right side and held a mirror so that when she rotated her head to the right she would be looking directly into it. What she would see was a reflection of the left side of the world which she had previously ignored. Would this make her suddenly realize that there's a whole left side to the world she had been ignoring so that she would turn to the left and look at it? If so we would have cured her neglect by merely using a

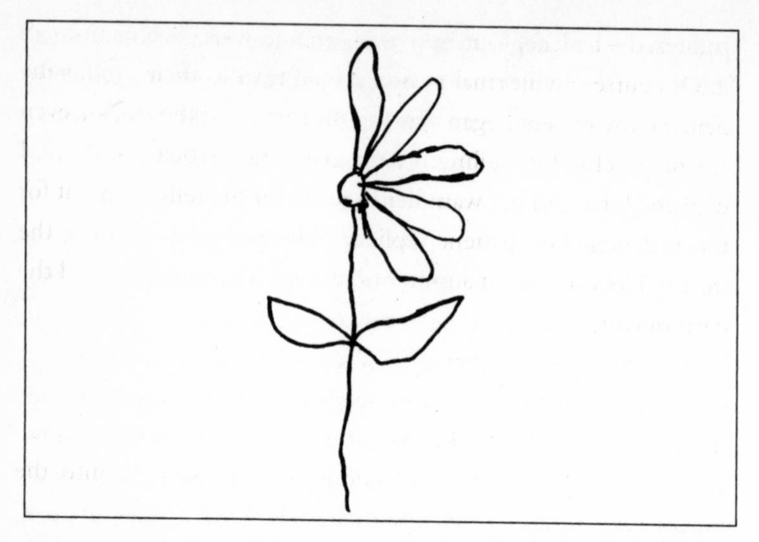

Figure 2.2 *Drawing made by a neglect patient. Notice that the left half of the flower is missing. Many neglect patients will also only draw half of the flower when drawing from memory—even with their eyes closed. This implies that the patient has also lost the ability to "scan" the left side of the internal mental picture of the flower.*

mirror! Or would her brain say (in effect) "Well, that's on my left, which doesn't really exist for me, so I'll continue to ignore it"?

The answer was, as often happens in science—neither! Before holding up the mirror I positioned John, my student, on her left side, holding a pen. I raised the mirror and asked the patient what she saw, what I was holding. The patient identified the mirror, saying that she could see her reflection in it and that it was cracked on the top—which it was.

She said she could also see John, and that he was holding a pen. I asked her to use her right hand—the hand that was not paralyzed—to reach out, take the pen and write her name with it. Of course, any normal person would turn to their left for the pen, but my patient began clawing the surface of the mirror, even reaching behind it, pulling my tie, grabbing my belt buckle ... I explained that I didn't want her to reach for the reflection but for the real pen. The patient replied: "The real pen is inside the mirror, Doctor," or, on another occasion: "The pen is behind the darn mirror, Doctor."

This is a problem that can be solved without difficulty by a three-year-old child. Even a chimpanzee doesn't confuse a mirror image for a real object. But the wise Mrs. D—in spite of seventy years of experience with reflections—reaches straight into the mirror. We call this "mirror agnosia" or "looking-glass syndrome" in honor of Alice, who actually walked into the mirror thinking it was a real world.

What causes mirror agnosia? I think what happens is the patient knows she's looking at a reflection, therefore the object is on her left. But because left doesn't exist in her universe the only possible explanation, however improbable, is that the object is inside the mirror. Remarkably, all her abstract knowledge about the laws of optics and mirrors is distorted to accommodate this strange new sensory world in which the patient finds herself trapped.[2] This isn't just some absent-minded or impulsive response to the mirror image: she actually recognizes the presence of the mirror and starts groping behind it, regarding it as an obstacle.

Another even more extraordinary disorder which is also caused by damage to the right parietal is denial or anosognosia. Remember, most patients with right parietal damage also have some damage to the internal capsule and so are completely paralyzed on the left side of the body. This is what is meant by a stroke, a complete paralysis of one side of the body. Most of them complain about this, as indeed they should, but a small percentage vehemently deny that their left arm is paralyzed, and some of these patients don't have any neglect. They keep saying their arm is moving correctly.

The fact that this behavior is usually seen only when the right parietal is damaged, but rarely when the left parietal is damaged, gives us a clue to what is happening. It seems that the denial syndrome has something to do with hemispheric specialization: the manner in which the two cerebral hemispheres deal with the external world, especially the manner in which they deal with discrepancies in sensory input and discrepancies in beliefs. Specifically, when confronted with a discrepancy, the left hemisphere's coping style is to smooth over it, pretend it doesn't exist and forge ahead. (Freudian defense mechanisms are an example of this.) The right hemisphere's coping style is the exact opposite. It is highly sensitive to discrepancies, so I call it the anomaly detector.

A patient with a right hemisphere stroke (left side paralyzed) sending a command to move his arm receives a visual feedback signal saying it is not moving, so there is a discrepancy. His right hemisphere is damaged, but his intact left hemisphere goes about its job of denial and confabulation, smoothing over the discrepancy and saying, all is fine, don't worry. On the other hand, if the left hemisphere is damaged and the right side is paralyzed, the right hemisphere is functioning as it should, so it notices the discrepancy between the motor command and the lack of visual feedback and recognizes the paralysis.[3] This was an outlandish idea but it's now been tested with brain imaging experiments and shown to be essentially correct.[4]

For a person to deny that he or she is paralyzed is quite bizarre, but seven or eight years ago we found something even more amazing. Some patients will also deny that *another* patient is paralyzed. I tell patient B, who is paralyzed, to move his arm. Patient B, of course, doesn't move, but if I ask my patient A who has anosognosia whether B moved his arm, A says yes, he did. Patient A is engaging in denial of another person's disabilities.[5]

At first this made no sense to me, but then I came across some studies by Giaccomo Rizzollati of experiments done on monkeys. It is well known that parts of the frontal lobes which are concerned with motor commands contain cells which fire when a monkey performs certain specific movements. One cell will fire when the monkey reaches out and grabs a peanut, another cell will fire when the monkey pulls something, yet another cell when the monkey pushes something. These are motor command neurons. Rizzollati found that some of these neurons will also fire when the monkey watches another monkey performing the same action. For example, a peanut-grabbing neuron which fires when the monkey grabs a peanut also fires when the monkey watches another monkey grab a peanut. The same thing happens in humans. This is quite extraordinary, because the visual image of somebody else grabbing a peanut is utterly different from the image of yourself grabbing a peanut—your brain must perform an internal mental transformation. Only then can that neuron fire in response both to its own movements *and* to another person making the same movements. Rizzollati calls these mirror neurons. Another name for them is monkey-see, monkey-do neurons, and these neurons are, I think, the ones that were damaged in our patients.

Consider what's involved in judging somebody else's movements. Maybe you need to do a virtual reality internal simulation of what that person is doing, and that may involve the activity of these very same neurons, these mirror neurons. So mirror neurons, instead of being some kind of curiosity, have important implications for understanding many aspects of human nature, such as interpreting somebody else's actions and intentions. We think it is this system of neurons that is damaged in some patients who have anosognosia. The patient can therefore no longer construct an internal model of somebody else's actions in order to judge whether that person is accurately carrying out a command or not.

I believe that these neurons may have played an important role in human evolution.[6] One of the hallmarks of our species is what we call culture. Culture depends crucially on imitation of parents and teachers, and the imitation of complex skills may require the participation of mirror neurons. I think that, somewhere around 50,000 years ago, maybe the mirror neurons system became sufficiently sophisticated that there was an explosive evolution of this ability to mime complex actions, in turn leading to cultural transmission of information, which is what characterizes us humans.

Mirror neurons also permit a sort of "virtual reality" simulation of other people's actions and intentions, which would explain why we humans are the "Machiavellian" primate—so good at constructing a "theory of other minds" in order to predict their behavior. This is indispensable for sophisticated social interactions, and some of our recent studies have shown that this system may be flawed in autistic children, which would explain their extreme social awkwardness.

But although the studies on patients are intriguing in themselves, our real agenda here is to understand how the *normal* brain works.[7] How does the activity of neurons generate the whole spectrum of abilities that we call human nature, whether it is body image or culture or language or abstract thinking? It is my belief that such a deeper understanding of the brain will have a profound impact not just on the sciences but on the humanities as well. Lofty questions about the mind are fascinating to ask—philosophers have been asking them for three millennia both in my native India and in the West—but it is only in the brain that we can eventually hope to find the answers.

3

The Artful Brain

In this chapter—the most speculative in the book—I consider one of the most ancient questions in philosophy, psychology and anthropology, namely, what is art? When Picasso said, "Art is the lie that reveals the truth," what exactly did he mean?

As we have seen, neuroscientists have made some headway in understanding the neural basis of psychological phenomena such as body image or visual perception. But can the same be said of art—given that art obviously originates in the brain?

In particular we can ask whether there are such things as artistic universals. There is obviously an enormous number of artistic styles across the world: Tibetan art, Classical Greek art, Renaissance art, Cubism, Expressionism, Impressionism, Indian art, pre-Columbian art, Dada … the list is endless. But despite this staggering diversity can we come up with some universal laws or principles that transcend these cultural boundaries and styles?

The question may seem meaningless to many social scientists; after all, science deals with *universal* principles whereas art is the ultimate celebration of human individuality and originality—the ultimate antidote to the homogenizing effects of science. There is some truth to this, of course, but even so I'd like to argue in this chapter that such universals do exist.

First, a note of caution. When I speak of "artistic universals" I am not denying the enormous role played by culture. Obviously, without culture there would be no different artistic styles—but neither does it follow that art is completely idiosyncratic and arbitrary, or that there are no universal laws.

To put it somewhat differently, let us assume that 90 percent of the variance seen in art is driven by cultural diversity or—more cynically—by just the auctioneer's hammer, and only 10 per cent by universal laws that are common to all brains. The culturally driven 90 percent is what most people already study—it's called art history. As a scientist, what I am interested in is the 10 per cent that is universal—not in the endless variations imposed by cultures. The advantage that scientists have today is that unlike philosophers we can now test our conjectures by directly studying the brain empirically. There's even a new name for this discipline. My colleague Semir Zeki calls it neuroaesthetics—just to annoy the philosophers.

I recently started reading about the history of ideas on art—especially Victorian reactions to Indian art—and it's a fascinating story. For example let's go to southern India and look at the famous Chola bronze of the goddess Parvathi dating back to the twelfth century (Figure 3.1). To Indian eyes, she is supposed to represent the very epitome of feminine sensuality, grace,

Figure 3.1 *Parvathi, consort of Lord Shiva; twelfth-century Chola dynasty (replica).*

poise, dignity, elegance: everything that's good about being a woman. And she's of course also very voluptuous.

But the Victorian Englishmen who first encountered these sculptures were appalled. Partly because they were prudish, but partly also because of just plain ignorance.

They complained that the breasts were far too big, the hips were too wide and the waist was too narrow. It didn't look anything like a real woman—it wasn't realistic—it was primitive art. And they said the same thing about the voluptuous nymphs of Kajuraho—even about Rajastani and Mogul miniature paintings. They said the paintings lacked perspective, that they were distorted.

The Victorians were unconsciously judging Indian art using the standards of Western art—especially classical Greek and Renaissance art, where realism is strongly emphasized.

But obviously this is a fallacy. Anyone today will tell you that art has nothing to do with realism. It is not about creating a replica of what's out there in the world. I can take a realistic photograph of my pet cat and no one would give me a penny for it. In fact, art is not about realism at all—it's the exact opposite. It involves deliberate hyperbole, exaggeration, even distortion, in order to create pleasing effects in the brain.

But obviously that can't be the whole story. You can't just take an image and randomly distort it and call it art. (Although in California, where I come from, many do!) The distortion has to be "lawful." The question then becomes, what kinds of distortion are effective? What are the laws?

I was sitting in a temple in India and in a whimsical frame of mind when I just jotted down what I think of as the universal laws of art, the ten laws of art which cut across cultural boundaries (see box).[1] The choice of 10 is arbitrary ... but it's a place to start.

The first law I call peak shift and to illustrate this I'll use a hypothetical example from animal behavior, from rat psychology.

Imagine you're training a rat to discriminate a square from a rectangle by giving it a piece of cheese every time it sees a particular rectangle. When it sees a square it receives nothing. Very soon it learns that the rectangle means food; it starts liking the rectangle—although a behaviorist wouldn't put it that way. Let's just say it starts going toward the rectangle because it prefers the rectangle to the square.

But if you take a longer, thinner rectangle and show it to the rat, it actually prefers the second rectangle to the first. This is because the rat is learning a rule—rectangularity. Longer and thinner equals more rectangular and, so far as the rat is concerned, the more rectangular, the better.

Professor Ramachandran's suggested 10 universal laws of art

1. Peak shift
2. Grouping
3. Contrast
4. Isolation
5. Perceptual problem solving
6. Symmetry
7. Abhorrence of coincidence/generic viewpoint
8. Repetition, rhythm and orderliness
9. Balance
10. Metaphor

And what has that to do with art?

Think about caricature. To produce a caricature of, say, Richard Nixon an artist must first ask: What's special about his face? What makes him different from other people? The artist will take the mathematical average, so to speak, of all male faces and subtract it from Nixon's face, leaving a big bulbous nose and shaggy eyebrows. These are then amplified to produce an image that looks even more like Nixon than Nixon himself. Skilled artists work this way to produce great portraiture;[2] take it a step further and you get caricature. It looks comical, but it still looks even more like Nixon than the original Nixon. So you're behaving exactly like that rat.

What has all this to do with the rest of art? Let's go back to the Chola bronze of Parvathi, where the same principle applies.

How does the artist convey the very epitome of feminine sensuality? He simply takes the average female form and subtracts the average male form—leaving big breasts, big hips and a narrow waist. And then amplifies the difference. The result is one anatomically incorrect but very sexy goddess.

But that's not all there is to it—what about dignity, poise, grace?

Here the Chola bronze artist has done something quite clever. There are some postures that are impossible for a male owing to the constraints imposed by pelvic anatomy, curvature of the lumbar spine and angle between the neck and shaft of the femur. I can't stand like that even if I want to. But a woman can do it effortlessly. So the artist visits an abstract space I call "posture space," subtracts the average male posture from the average female and then exaggerates it. Doing this produces the elegant triple flexion—or tribhanga—pose, where the head is tilted one way, the body is tilted exactly the opposite way, and the hips again the other way. And again the viewer's reaction is not that the figure is anatomically inappropriate because nobody can stand like that. What the viewer sees is a gorgeous, beautiful, celestial goddess. This extremely evocative image is an example of the peak shift principle in Indian art.

So much for faces and caricatures and bodies and Chola bronzes. But what about the rest of art? What about abstract art, semi-abstract art, Impressionism, Cubism? What about Picasso, Van Gogh, Monet, Henry Moore? How can my ideas even begin to explain the appeal of some of those artistic styles?

To answer this question, we need to look at evidence from

ethology, especially the work of Niko Tinbergen at Oxford more than fifty years ago, who was doing some very elegant experiments on herring-gull chicks.

As soon as the herring-gull chick hatches, it sees its mother's long yellow beak with a red spot on it. It starts pecking at the red spot, begging for food. The mother then regurgitates half-digested food into the chick's gaping mouth, the chick swallows the food and is happy. Tinbergen asked himself: "How does the chick recognize its mother? Why doesn't it beg for food from a person who is passing by or from a pig?"

And he found that you don't need a mother. A hatchling would react in exactly the same way to a disembodied beak with no mother attached.

Why does a chick think a scientist waving a beak is a mother seagull? Well, the goal of vision is to do as little processing or computation as is necessary for the job on hand, in this case for recognizing mother. And through millions of years of evolution, the chick has acquired the wisdom that this long thing with a red spot always has a mother attached to it, rather than a mutant pig or a malicious ethologist. So it can take advantage of the statistical redundancy in nature and assume: "Long yellow thing with a red spot equals mother," thereby simplifying the processing and saving a lot of computational labor.

That seems fair enough. But what Tinbergen found next is that he didn't need even a beak. He took a long yellow stick with three red stripes, which looked nothing like a beak—and that's important—and the chicks pecked at the stick even more than they would have pecked at a real beak. They preferred it to a real beak, even though it didn't resemble a beak. Tinbergen had stumbled on a superbeak—an ultrabeak. So the chick's brain goes: "Wow—what a sexy beak!"

Why does this happen? We don't know exactly, but obviously there are neural circuits in the visual pathways of the chick's brain that are specialized to detect a beak as soon as the chick hatches. They fire upon seeing the beak. Perhaps because of the way they are wired up, they may actually respond more powerfully to the stick with three stripes than to a real beak. Maybe the neurons' receptive field embodies a rule such as "the more red contour the better." And so even though the stick doesn't look like a beak—maybe not even to the chick—this strange object is actually more effective in driving beak detectors than a real beak. And a message from this beak-detecting neuron goes to the emotional limbic centers in the chick's brain, giving it a big jolt and the message: "Here is a superbeak." The chick is absolutely mesmerized.

All of which brings me to my punch-line about art. If herring-gulls had an art gallery, they would hang a long stick with three red stripes on the wall; they would worship it, pay millions of dollars for it, call it a Picasso, but not understand why—why they are mesmerized by this thing even though it doesn't resemble anything. That's all any art lover is doing when buying contemporary art: behaving exactly like those gull chicks.

In other words human artists through trial and error, through intuition, through genius, have discovered the figural primitives of our perceptual grammar. They are tapping into these and creating for the human brain the equivalent of the long stick with three stripes. And what emerges is a Henry Moore or a Picasso.

The advantage with these ideas is that they can be tested experimentally. It is possible to record from cells in the fusiform gyrus of the brain that respond powerfully to individual faces. Some of them will fire only to a particular view of a face, but higher up are found neurons each of which will respond to *any* view (profile vs. full frontal) of a given face. And I predict that if you present a monkey with a Cubist portrait of a monkey's face—two different views of a monkey's face superimposed in the same location in the visual field—then that cell in the monkey's brain will be hyperactivated just as a long stick with three stripes hyperactivates the beak-detecting neurons in the chick's brain. So what we have here is a neural explanation for Picasso—for Cubism.[3]

I've discussed one of my universal laws of art so far—peak shift and the idea of ultra-normal stimuli—and have borrowed insights from ethology, neurophysiology and rat psychology to account for why people like non-realistic art.[4,5]

The second law is more familiar. It's called grouping.

Most of us are familiar with puzzle pictures, such as Richard Gregory's Dalmatian dog. At first sight you see nothing but a bunch of splotches, but you can sense your visual brain trying to solve a perceptual problem, trying to make sense of this chaos. And then after 30 or 40 seconds suddenly everything clicks in place and you group all the correct fragments together to see a Dalmatian dog (Figure 3.2).

You can almost sense your brain groping for a solution to the perceptual riddle and as soon as you successfully group the correct fragments together to see the object, what I suggest is that a message is sent from the visual centers of the brain to the limbic-emotional centers of the brain, giving it a jolt and saying: "A-ha, there is an object—a dog," or "A-ha, there is a face."

The Dalmatian example is very important because it reminds us

Figure 3.2 *Gregory's Dalmatian dog (photo by Ron James).*

that vision is an extraordinarily complex and sophisticated process. Even looking at a simple scene involves a complex hierarchy, a stage-by-stage processing. At each stage in the hierarchy of processing, when a partial solution is achieved—when a part of the dog is identified—there is a reward signal "a-ha," a partial "a-ha," and a small bias is sent back to earlier stages to facilitate the further binding of the features of the dog. And through such progressive bootstrapping the final dog clicks in place to create the final big "A-HA!" Vision has much more in common with problem solving—like a twenty-questions game—than we usually realize.

The grouping principle is widely used in both Indian and in Western art—and even in fashion design. For example, you go shopping and pick out a scarf with red splotches on it. Then you look for a skirt which has also got some red splotches on it.

Why? Is it just hype, just marketing, or is it telling you something very deep about how the brain is organized? I believe it is telling you something very deep, something to do with the way the brain evolved.

Vision evolved mainly to discover objects and to defeat camouflage. You don't realize this when you look around you and you see clearly defined objects, but imagine your primate ancestors scurrying up in the treetops trying to detect a lion seen behind fluttering green foliage. What you get inside the eyeball on the retina is just a mass of yellow lion fragments obscured by all the leaves. But the visual system of the brain "knows" that the likelihood that all these different yellow fragments being exactly the same yellow simply by chance is zero. They must all belong to one object. It links them together, decides it's a lion (based on the overall shape) and sends a big "a-ha" signal to the limbic system telling you to run.

Arousal and attention culminate in titillating the limbic system. Such "a-has" are created, I maintain, at every stage in the visual hierarchy as partial object-like entities are discovered that draw our interest and attention. What an artist tries to do is to generate as many of these "a-ha" signals in as many visual areas as possible by more optimally exciting these areas with painting or sculpture than could be achieved with natural visual scenes or realistic images. Not a bad definition of art, if you think about it.

That brings me to my third law—the law of perceptual problem solving or visual peek-a-boo.

As anyone knows, a nude seen behind a diaphanous veil is much more alluring and tantalizing than a full-color *Playboy* photo or a Chippendale pin-up. Why? (This question was first raised by the Indian philosopher Abhinavagupta in the tenth century.) After all, the pin-up is much richer in information and should excite many more neurons.

As I have said, our brains evolved in highly camouflaged environments. Imagine you are chasing your mate through dense fog. Then you want every stage in the process—every partial glimpse of her—to be pleasing enough to prompt further visual search—so you don't give up the search prematurely in frustration. In other words, the wiring of your visual centers to your emotional centers ensures that the very act of searching for the solution is pleasing, just as struggling with a jigsaw puzzle is pleasing long before the final "a-ha." Once again it's about generating as many "a-has" in your brain as possible.[6] Art may be thought of as a form of visual foreplay before the climax.

We have discussed three laws so far: peak shift, grouping and perceptual problem solving. Before I go any further I'd like to emphasise that looking for universal laws of aesthetics does not negate the enormous role of culture, nor the genius and originality of an individual artist. Even if the laws are universal, which particular law (or combination of them) an artist chooses to deploy depends entirely on his or her genius and intuition. Thus, while Rodin and Henry Moore were mainly tapping into "form," Van Gogh and Monet were mainly introducing peak shifts in an abstract "color space"—brain maps in which adjacent points in color space rather than Cartesian space are mapped adjacently. Hence the effectiveness of artificially heightened "non-realistic" colors of their sunflowers or water lilies. These two artists also deliberately blurred the outlines to avoid distracting attention from the colors where it was needed most. Other artists may choose to emphasise even more abstract attributes such as shading or illumination (Vermeer).

And that brings us to my fourth law—the law of isolation or understatement.

A simple outline doodle of a nude by Picasso, Rodin or Klimt can be much more evocative than a full-color pin-up photo. Similarly the cartoon-like outline drawings of bulls in the Lascaux Caves are much more powerful and evocative of the animal than a *National Geographic* photograph of a bull. Hence the famous aphorism: "Less is more."

But why should this be so? Isn't it the exact opposite of the first law, the idea of hyperbole, of trying to excite as many "a-has" as possible? A pin-up or a Page Three photo has, after all, much more information. It's going to excite many more areas in the brain, many more neurons, so why isn't it more beautiful?[7]

The answer to this paradox lies in another visual phenomenon: "attention." It is well known that there cannot be two overlapping patterns of neural activity simultaneously. Even though the human brain contains a hundred billion nerve cells, no two patterns may overlap. In other words, there is a bottleneck of attention. Attentional resources may be allocated to only one entity at a time.

The main information about the sinuous, soft contours of a Page Three girl is conveyed by her outline. Her skin tone, hair color, etc. are irrelevant to her beauty as a nude. All this irrelevant information clutters the picture and distracts attention from where it needs critically to be directed—to her contours and outlines. By omitting such irrelevant information from a doodle or sketch the artist is saving your brain a lot of trouble. And this is especially true if the artist has also added some peak shifts to the outline to create an "ultra nude" or a "super nude."

This theory can be tested by doing brain imaging experiments comparing neural responses to outline sketches and caricatures versus full-color photos. But there is also some striking neurological evidence from children with autism. Some of these children have what is known as the savant syndrome. Even though they are retarded in many respects, they have one preserved island of extraordinary talent.

For example, a seven-year-old autistic child, Nadia, had exceptional artistic skills. She was quite retarded mentally, could barely talk, yet she could produce the most amazing drawings of horses and roosters and other animals. A horse drawn by Nadia would almost leap out at you from the canvas (Figure 3.3 left). Contrast this with the lifeless, two-dimensional, tadpole-like sketches drawn by most normal eight- or nine-year-olds (right)—or even a very good one by Leonardo da Vinci (center).

So we have another paradox. How can this retarded child produce a drawing that is so incredibly beautiful? The answer, I maintain, is the principle of isolation.

In Nadia, perhaps many or even most of her brain modules are damaged because of her autism, but there is a spared island of cortical tissue in the right parietal. So her brain spontaneously allocates all her attentional resources to the one module that's still functioning, her right parietal. The right parietal is the part of the brain concerned with our sense of artistic proportion. We know this because when it's damaged in an adult, artistic sense is lost. Stroke patients with right parietal damage produce drawings that are often excessively detailed but lack the vital essence of the

Figure 3.3 *(a) A drawing of a horse made by Nadia, the autistic savant, when she was five years old. (b) A horse drawn by Leonardo da Vinci. (c) A drawing of a horse by a normal eight-year-old. Notice Nadia's drawing is vastly superior to that of the normal eight-year-old and almost as good as (or perhaps better than!) Leonardo's horse. (a) and (c) reprinted from* Nadia, *by Lorna Selfe, with permission from Academic Press (New York).*

picture they are trying to depict. They have lost their sense of artistic proportion. Nadia, since everything else is damaged in her brain, spontaneously allocates all her attention to the right parietal—so she has a hyperfunctioning art module in her brain which is responsible for her beautiful renderings of horses and roosters. What most of us "normals" have to learn to do through years of training—ignoring irrelevant variables—she does effortlessly. Consistent with this idea, Nadia lost her artistic sense once she grew up and improved her language skills.

Another example is equally striking. Steve Miller, of the University of California, has studied patients who start developing rapidly progressing dementia in middle age, a form of dementia called the fronto-temporal dementia. This affects the frontal and temporal lobes, but spares the parietal lobe. Some of these patients suddenly start producing the most amazingly beautiful paintings and drawings, even though they had no artistic talent before the onset of their dementia. Again, the isolation principle is at work. With all other modules in the brain not working, the patient develops a hyper-functioning right parietal. There are even reports from Alan Snyder in Australia that it is possible to unleash such hidden talents by temporarily paralyzing parts of the brain in normal volunteers. If his findings are confirmed, it will truly be a brave new world.

That brings me to another question: why do humans even bother creating and viewing art?[8] I've already hinted at some possible answers but let me spell them out more explicitly. There are at least four possibilities—none mutually exclusive.

First, it is possible that once laws of aesthetics have evolved (for reasons such as discovering, attending to and identifying objects) then they may be artificially hyperstimulated even though such titillation serves no direct adaptive purpose, just as saccharin tastes "hypersweet" even though it provides zero energy and zero nutrition.

Second, as suggested by Miller, artistic skill may be an index of skillful eye–hand coordination and, therefore, an advertisement of good genes for attracting potential mates (the "come and see my etchings" theory). This is a clever idea that I don't find convincing. It doesn't explain why the so-called "index" takes the particular form that it does: art. After all, few women—not even feminists!—find the ability to knit or embroider attractive in a man, even though these demand excellent eye–hand coordination. My point is, why not use a much more straightforward "index" such as proficiency in archery or javelin throwing (which, to be sure, *are* attractive in a man)?

Third, there is Steve Pinker's idea that people acquire art as a status symbol to advertise their wealth: the "I own a Picasso, so help me spread our genes together" theory. Anyone who has been to a cocktail reception at an art gallery knows there's some truth to this.

Fourth—the idea I favor—art may have evolved as a form of virtual reality simulation. When you imagine something—as when rehearsing a forthcoming bison hunt or amorous encounter—many of the same brain circuits are activated as when you really do something. This allows you to practice scenarios in an internal simulation without incurring the energy cost or risks of a real rehearsal.

But there are obvious limits. Evolution has seen to it that our imagery—internal simulation—isn't perfect. A hominid with mutations that enabled it to perfectly imagine a feast instead of having one, or imagine orgasms instead of pursuing mates, is unlikely to spread its genes. This limitation in our ability to create internal simulations may have been even more apparent in our ancestors. For this reason they may have created real images ("art") as "props" to rehearse real bison hunts or to instruct their children. If so, we could regard art as Nature's own "virtual reality" (just as my mirror box allows patients to actually see their phantom arm and move it—whereas they couldn't do so just using imagination).

Limitations of space prevent the discussion of all my other laws in detail, but I will mention the last on my list. In many ways it is the most important, yet the most elusive: visual metaphor. A metaphor in literature juxtaposes two seemingly unrelated things to highlight certain important aspects of one of them (as when the Indian poet Rabindranath Tagore referred to the Taj Mahal as "A teardrop on the cheek of time"). The same thing is possible in visual art. For example, the multiple arms on the Chola bronze of the dancing Shiva or Nataraja (Figure 3.4) are not meant to be taken literally, as they were by the Victorian art critic Sir George Birdwood, who called it a multi-armed monstrosity. (Funnily enough, he didn't think that angels sprouting wings were monstrosities—although I can tell you as a medical man that to possess multiple arms is anatomically possible, but wings on scapulae are not!)

The multiple arms are meant to symbolize multiple divine attributes of God and the ring of fire that Nataraja dances in—indeed his dance itself—is a metaphor of the dance of the cosmos and of the cyclical nature of creation and destruction, an idea championed by the late Fred Hoyle. Most great works of art—be they Western or Indian—are pregnant with metaphor and have many layers of meaning.[9]

Everyone knows that metaphors are important, yet we have no idea why. Why not just say "Juliet is radiant and warm" instead of saying "Juliet is the sun"? What is the neural basis for metaphor? We don't know, but I will attempt some answers in chapter 4.

Many social scientists feel rather deflated when informed that beauty, charity, piety and love are the result of the activity of neurons in the brain, but their disappointment is based on the false assumption that to explain a complex phenomenon in terms of its component parts ("reductionism") is to explain it away. To understand why this is a fallacy imagine it's the twenty-second century and I am a neuroscientist watching you and your partner

Figure 3.4 *Nataraja or dancing Shiva. Chola dynasty copper alloy, thirteenth century.*

(Esmeralda) making love. I scan Esmeralda's brain and tell you everything that's going on in it when she is in love with you and is making love to you. I tell you about activity in her septum, in her hypothalamic nuclei, and how certain peptides are released along with the affiliation hormone prolactin, etc. You might then turn to her and say, "You mean that's all there is to it? Your love isn't real? It's all just chemicals?" To which Esmeralda should respond, "On the contrary, all this brain activity provides hard evidence that I *do* love you, that I'm not just faking it. It should increase your confidence in the reality of my love." And the same argument holds for art or piety or wit.

Do these laws of neuroaesthetics encompass everything there is to know about art? Of course not; I have barely scratched the surface. But I hope that these laws have provided some hints about the general form of a future theory of art, and about how a neuroscientist might try to approach the problem.

The solution to the problem of aesthetics, I believe, lies in a more thorough understanding of the connections between the thirty visual centers in the brain and the emotional limbic structures (and of the internal logic and evolutionary rationale that drives them). Once we have achieved a clear understanding of these connections, we will be closer to bridging the huge gulf that separates C. P. Snow's two cultures—science on the one hand and arts, philosophy and humanities on the other.

We could be at the dawning of a new age where specialization becomes old-fashioned and a twenty-first-century version of the Renaissance man is born.

4

Purple Numbers and Sharp Cheese

You know my method, Watson. It is founded upon the observation of trifles.

SHERLOCK HOLMES

In the nineteenth century the Victorian scientist Francis Galton, who was a cousin of Charles Darwin, noticed something very odd. He found that certain people in the population who were otherwise perfectly normal would experience a specific color every time they heard a specific tone. For example, C sharp might be red, F sharp might be blue, another tone might be indigo. He called this curious mingling of the senses synesthesia. Some of these people also see colors when they see numbers. Every time they see a black number, say the number five, printed on a white page (or a white five on a black page, for that matter), they would see it tinged, say, red. Six might be green, seven indigo, eight yellow and so on. Galton also asserted that this condition runs in families, something which Simon Baron-Cohen in Cambridge has recently confirmed.

It is fair to say that, even though synesthesia has been known about for over a hundred years, it has been by and large regarded as a curiosity and has never made it into mainstream neuroscience and psychology. But such "anomalies" (to use Thomas Kuhn's phrase) can be extremely important in science. Of course, most anomalies turn out to be false alarms, such as telepathy, spoon bending or cold fusion, but if you pick the right one you can completely change the direction of research in a field and generate a scientific revolution.

But first let's look at the most common explanations that have been proposed to account for synesthesia.

There are four. The first explanation is the most obvious: that these people are just crazy. It's the common reaction of scientists: when something does not fit the accepted "big picture" it just gets brushed under the carpet. The second explanation is that they've been on drugs. This is not an entirely inappropriate criticism because synesthesia is more common among people who use LSD, but in my view that makes it more interesting, not less so. Why should some chemicals cause synesthesia (if indeed they do)?

The third idea is that synesthetes are just recalling childhood memories. For example, maybe they saw refrigerator magnets where five was red and six was blue and seven was green, and for some reason they're stuck with these memories. This has never made much sense to me because if it were true why should the condition run in families? (Unless the same magnets are passed down, or the propensity to play with magnets runs in families ...) The fourth explanation is more subtle and it invokes sensory metaphors. Our everyday language is replete with synesthetic metaphors, cross-sensory metaphors: for example "Cheddar cheese is sharp." Well, cheese isn't sharp, it's soft. What we mean is it tastes sharp; "sharp" is a metaphor. But this argument is circular—why use a tactile adjective, sharp, for a taste sensation?

In science, one mystery cannot be explained with another mystery. Saying that synesthesia is just a metaphor explains nothing because we don't know what a metaphor is or how it's represented in the brain. Indeed, I'd suggest the very opposite, that synesthesia is a sensory phenomenon whose neural basis can be discovered in the brain and that in turn can provide an experimental foothold for understanding more elusive aspects of the mind, such as metaphor.

Why has synesthesia been ignored for so long? There's an important lesson here in the history of science. I think in general it is fair to say that for a curious phenomenon, an anomaly, to make it into mainstream science and have an impact, it has to fullfil three criteria. First, it must be a demonstrably real phenomenon ... it has to be reliably repeatable under controlled conditions. Second, there must be a candidate mechanism that explains the phenomenon in terms of previously known principles. And third, it has to have significant implications beyond the phenomenon itself. Take, for example, telepathy. Telepathy has vastly significant implications if real, so the third criterion is fulfilled, but the first criterion is not fulfilled; it's not reliably repeatable. We don't even know if it's a real phenomenon.

Indeed, the more you measure it the smaller it becomes and that's always a bad sign. Another example is bacterial transformation. Some years ago it was discovered that incubating one species of the bacteria pneumococcus with another species resulted in the second species becoming transformed into the first. In fact, the transformation could be induced simply by extracting the chemical we would today call the bacteria's DNA. This was published in a prestigious journal and was reliably repeatable. Yet people ignored it because nobody could think of a candidate mechanism. How could you possibly encode heredity in a chemical? Then Watson and Crick described the double-helical structure of DNA and cracked the genetic code. Once that happened, the scientific community perked up and recognized the importance of bacterial transformation.

I have tried to do something similar with synesthesia. First of all, I will try to show it's real, not bogus. Second, I will suggest candidate mechanisms, what is going on in the brain. And third, I will argue that synesthesia has very broad implications. It might tell us about things like metaphor and how language evolved in the brain, maybe even the emergence of abstract thought that we human beings are very good at.

In an attempt to show that synesthesia is a real phenomenon, my colleagues and I essentially developed a clinical test to identify closet synesthetes. First, we found two synesthetes who saw numbers as colored; in their case 5 was green and 2 was red. We produced a computerized display which had a random jumble of 5s on the screen and embedded among these 5s a number of 2s, arranged to form a geometric shape. A non-synesthete would take as long as twenty seconds to see the arrangement of all the

Figure 4.1 *A "clinical test" for synesthesia. The display consists of 2s embedded in a matrix of randomly placed 5s. Non-synesthetes find it very hard to discern the embedded shape (in this case a triangle). Synesthetes who see the numbers as colored can detect the triangle much more easily. (Depicted schematically in this figure; compare with Figure 1.8.)*

2s (Figure 1.8), but the two synesthetes immediately or very quickly saw the shape formed by the red 2s against the background of contrasting green 5s (shown schematically in Figure 4.1). They are obviously not crazy: how can someone crazy outperform normals? And the effect must be sensory rather than based on memory, or they wouldn't be able to literally see the geometric shape. Using this and other similar tests, we have discovered that synesthesia is much more common than has been assumed in the past. In fact, we found that one in two hundred people is a synesthete.

What causes synesthesia? In 1999, my student Ed Hubbard and I were looking at a structure called the fusiform gyrus in the temporal lobes. The fusiform gyrus contains the color area V4 which was described by Semir Zeki (Zeki and Marini, 1998). This is the area which processes color information, but we were struck by the fact that the number area of the brain, which represents visual numbers as shown by brain imaging studies, is right next to it, almost touching the color area of the brain (Figure 4.2). This is an unlikely coincidence; the most common type of synesthesia is number/color synesthesia *and* the number area and color areas are right next to each other in the same part of the brain. It seemed likely that these people had some accidental cross-talk, or cross-wiring, just as in my phantom limb patients (see chapter 1). The difference here is that it happens not because of amputation but because of some genetic change in the brain. Imaging experiments on people with synesthesia (Figure 4.3) suggest that showing black and white numbers to a synesthete produces activation in the color area of the fusiform gyrus.[1]

Further evidence for this "cross-activation" theory came from a most unexpected observation. We recently came across a man who was partially color-blind but had full-blown synesthesia. Because of a deficiency in his cone pigments (in the retina) he couldn't see the full range of colors in the world. Yet when looking at numbers he could see colors that he could never experience otherwise. He referred to them charmingly as "Martian colours." We suggest that this occurs because, even though

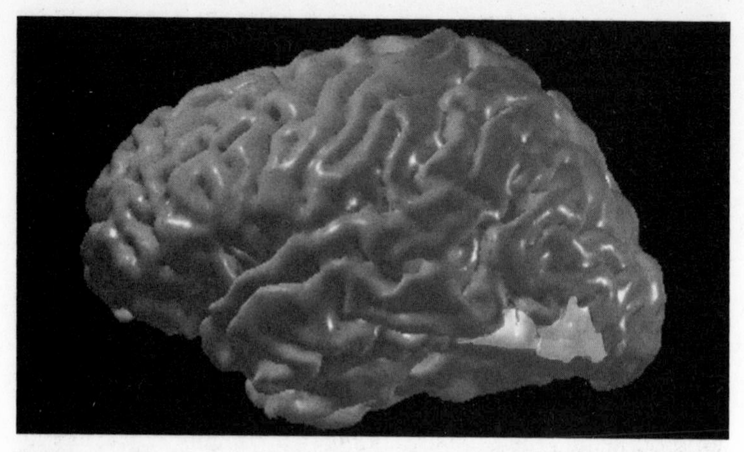

Figure 4.2 *Shows that the color area V4 and the so-called "number-grapheme area" are close to each other in the fusiform gyrus of the temporal lobes. A "cross-activation" between these adjacent areas may provide a neural substrate for synesthesia. Area V4 is indicated by diagonal lines while the number-grapheme area is indicated by cross-hatching.*

his receptors are deficient in the eye, the color areas in his brain are normal and can be accessed indirectly through cross-activation by numbers. The observation provides strong evidence against the memory association hypothesis: how could he remember something he had never actually seen?

Intriguingly, in some synesthetes even an invisible number can evoke color. If a number (say, 5) is presented off to one side and flanked by two other numbers, which we call distractors, then normal people find it hard to discern the middle number—an effect called crowding. This isn't caused by a decline of visual acuity in peripheral vision because the same number is readily seen if the two flanking distractors are removed (Figure 4.4).

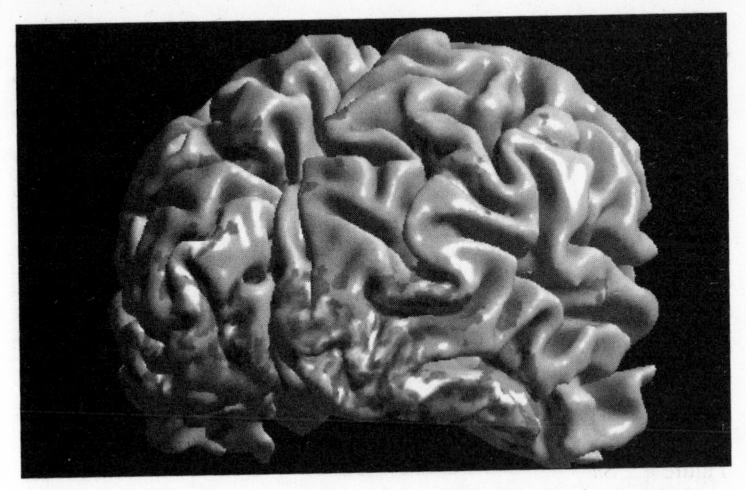

Figure 4.3 *Rear view of a synesthete's brain; fMR (functional magnetic resonance) image showing high activity in V4—a color processing area—as the subject looks at white numbers on a grey background. This area is not active in normal people viewing the same figures.*

Crowding occurs because the flanking numbers distract attention from the middle number.

But a synesthete, unable to discern the number itself, will still identify a 5 "because it looks red"—implying that even a number that is not consciously visible can nevertheless evoke color! I suggest that the cross-activation occurs before the stage where the number reaches conscious awareness and the color evoked is then relayed to higher centers in the brain where it is consciously perceived and used to deduce intellectually what the number must have been.[2] This phenomenon bears an uncanny resemblance to what we called blindsight in chapter 2. It would also explain why many synesthetes actually use their colors as a

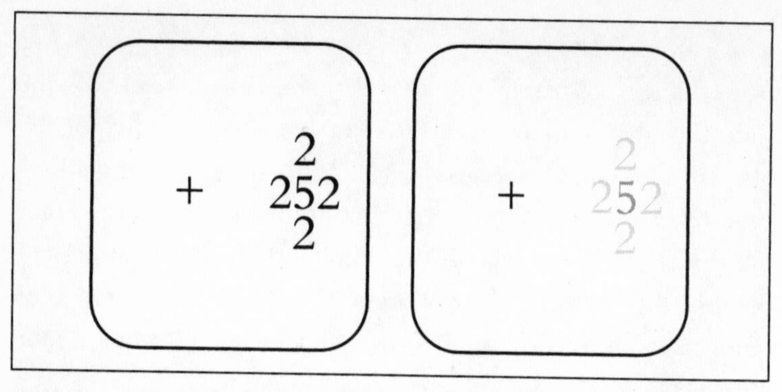

Figure 4.4 *"Invisible numbers." When a normal person stares at the central fixation sign (here a plus sign, +), a single digit off to one side is easy to see with peripheral vision (left). But, if the number is surrounded by other flanking numbers it appears indiscernible to the average person. In contrast a synesthete could deduce the central number by the color it evokes.*

mnemonic aid, e.g., for learning phone numbers and musical scales.

Why does this cross-wiring or cross-activation occur? The fact that it runs in families suggests that there is a gene, or set of genes, involved. What might this bad gene be doing? One possibility is that we are all born with excess connections in the brain. In the fetus there are many redundant connections which get pruned away to produce the modular architecture characteristic of the adult brain. What I think has happened in these people is that the "pruning" gene is defective, which has resulted in cross-activation between areas of the brain.[3] Or perhaps some kind of chemical imbalance has caused cross-activation between adjacent parts of the brain that are normally only loosely connected.

What we found next was even more amazing. We showed our two subjects Roman numbers V and VI instead of the Indian/Arabic numbers 5 and 6. They knew it was a five or six, but saw no color. This result is very important because it demonstrates that it is not the numerical concept that drives the color but the visual appearance of the number. This fits my argument because the fusiform gyrus represents the *visual* appearance of numbers and letters, not the abstract concept of sequence or ordinality.[4]

We don't know where in the brain the abstract idea of number is represented, but a good guess is the angular gyrus in the left hemisphere. When that area is damaged, patients can no longer do arithmetic even though they can see and identify numbers correctly. They are fluent in conversation, they remain intelligent, but cannot do even simple calculations such as seventeen minus three. This would indicate that abstract number concepts are represented in the angular gyrus while the fusiform gyrus deals with the visual appearance of a number.

But not all synesthetes are created equal. We soon ran into others in whom not merely numbers but even days of the week and even months of the year evoke colors: Monday is red, Tuesday is indigo, December is yellow. No wonder people thought they were crazy! What days, months and numbers have in common, though, is the abstract idea of sequence or ordinality, which I believe is represented higher up in the temporal parietal occipital (TPO) junction, in the vicinity of the angular gyrus (Figure 4.5). It should by now come as no surprise that the next color area in the color processing hierarchy is higher up in the general vicinity of the TPO junction, not far from the angular

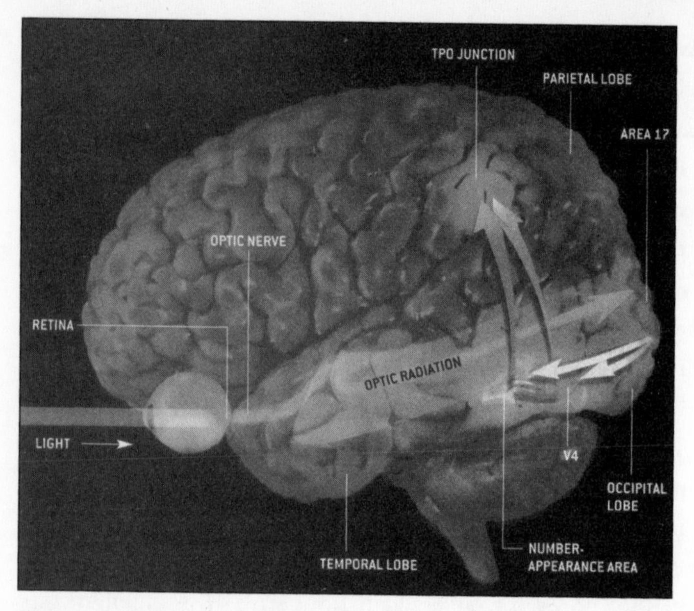

Figure 4.5 *Stages of number and color processing in the human brain. (Illustration by Carol Donner.)*

gyrus. So I think that in these people, those who see colors in days and months, the cross-wiring is higher up in the angular gyrus. For this reason, I call them higher synesthetes. In summary, if the faulty gene is selectively expressed in the fusiform gyrus, at an earlier stage in processing, the result is a lower synesthete driven by visual appearance. If the gene is expressed selectively higher up, in the vicinity of the angular gyrus, the result is a higher synesthete driven by numerical concept rather than visual appearance. Such selective gene expression can occur as a result of transcription factors.[5]

One in two hundred people has this completely useless peculiarity of seeing colored numbers—why did this gene survive? I would suggest it's a bit like sickle-cell anemia—there's a hidden agenda. These genes are doing something else important.[6]

One of the odd facts about synesthesia, which has been known, and ignored, for a long time, is that it is seven times more common among artists, poets, novelists—in other words, flaky types! Is this because artists are crazy? Or just willing to report their experiences unselfconsciously? Or maybe even trying to attract attention to themselves? (It's sexy to be a synesthete, given that many eminent artists are or were.) But I'd propose a very different view. What artists, poets and novelists all have in common is their skill at forming metaphors, linking seemingly unrelated concepts in their brain, as when Macbeth said "Out, out brief candle," talking about life. But why call it a candle? Is it because life is like a long white thing? Obviously not. Metaphors are not to be taken literally (except by schizophrenics, which is another story altogether). But in some ways life *is* like a candle: it's ephemeral, it can be snuffed out, it illumines only very briefly. Our brains make all the right links, and Shakespeare, of course, was a master at doing this. Now, let's make one further assumption—that this "cross-activation" or "hyperconnectivity" gene is expressed more diffusely throughout the brain, and not just in the fusiform or in the angular. As we have seen, if the gene is expressed in the fusiform you get a lower synesthete, and if expressed in the angular gyrus/TPO junction you get a higher synesthete. But if it's expressed everywhere there is greater hyperconnectivity throughout the brain, making that person more prone to metaphor, the ability to link seemingly unrelated things. (After all, even so-called abstract concepts are also represented in brain maps.) This may seem counter-intuitive, but just think of something like a number. There is nothing more abstract than a number. Five pigs, five donkeys, five chairs, even five tones—all very different, but with fiveness in common. That fiveness is represented in a fairly small region, the angular gyrus. So it's possible that other high-level concepts are also represented in brain maps and that artistic people, with their excess connections, can make these associations much more fluidly and effortlessly than less gifted people.[7]

So far we have shown that synesthesia is a genuine sensory effect and proposed a candidate mechanism, so satisfying the first two criteria outlined above for its entry into mainstream science. All that remains is to show that synesthesia is not just some oddity—something weird—but has implications beyond the confines of a narrow speciality. In my view, synesthesia is far more than just a quirk in some people's brains. In fact, most of us are synesthetes, as I will now demonstrate. Imagine that in front of you is a bulbous amoeboid shape on which are many undulating curves. And right next to it imagine a jagged shape—like a piece of shattered glass with sharp jagged edges (Figure 4.6). These shapes are the first two letters of the Martian alphabet. One of these shapes is kiki and the other is booba, and you have to decide which is which. Look at the figure now and decide which is kiki. In experiments, 98 percent of people say the jagged shape is kiki and the bulbous amoeboid shape is booba. If you are among them you're a synesthete. Let me explain. Look at the letter kiki and compare it with the sound "kiki." They both share one property: the kiki visual shape has a sharp inflexion and the sound "kiki" represented in your auditory cortex, in the

Figure 4.6 *If asked which of these two abstract shapes is "booba" and which "kiki", 95–98 percent of respondents pick the blob as booba and the jagged shape as kiki. This is also true for non-English-speaking Tamillians for whom the shapes bear no resemblance to visual shapes of the Tamil alphabet corresponding to B or K. The effect demonstrates the brain's ability to engage in cross-modal abstraction of properties such as jaggedness or curviness. Our preliminary results suggest that the effect is compromised in patients with left angular gyrus lesions who also have difficulties with metaphor.*

hearing centers of your brain, also has a sharp sudden inflection. Your brain performs a cross-modal synesthetic abstraction, recognizing that common property of jaggedness, extracting it, and so reaching the conclusion that they are both kiki.

(Interestingly, Tamillians, who don't speak or write English, produce the same results, so this phenomenon is unrelated to the jagged shape resembling the visual appearance of the letter K. Other shapes can also be paired with sounds in this manner: for example, if you show a blurred or smudged line and a sawtooth and ask people which is "rrrrrr" and which is "shhhhh" they spontaneously pair the former with "shhhhh" and the latter with "rrrrrr.")

We have tried the booba/kiki experiment on patients who have a very small lesion in the angular gyrus of the left hemisphere. Unlike you and me, they make random shape–sound associations. They cannot do this cross-modal abstraction even though they're fluent in conversation, they're intelligent, and seem quite normal in other respects. This makes perfect sense because the angular gyrus (Figure 1.3) is strategically located at the crossroads between the parietal lobe (concerned with touch and proprioception), the temporal lobe (concerned with hearing) and the occipital lobe (concerned with vision). So it is strategically placed to allow a convergence of different sense modalities to create abstract, modality-free representations of things around us. Logically, the jagged shape and the sound "kiki" have nothing in common: the shape comprises photons hitting the retina in parallel, the sound is a sharp air disturbance hitting the hair cells of the inner ear sequentially. But the brain abstracts the common denominator—the property of jaggedness. Here in the angular gyrus are the rudimentary beginnings of the property we call abstraction that we human beings excel in.

Why did this ability evolve in humans in the first place? Why cross-modal abstraction? If we compare the brains of lower mammals, monkeys, great apes and humans, we find a progressive enlargement of the TPO junction and angular gyrus, an almost explosive development. And especially so in humans. I think this ability initially evolved to help us survive in treetops, grabbing handholds, jumping from branch to branch. To do this it is necessary to adjust the angle of the arm and the fingers so that the proprioceptive map (signaled by receptors in muscles and joints) matches the horizontality of a branch's visual appearance—the horizontal array of photons—which is why the angular gyrus became larger and larger. But once this ability to engage in cross-modal abstraction was developed, that structure in turn became an exaptation for the other types of abstraction that modern humans excel in, be it metaphor or any other type of abstraction.[8] Such opportunistic hijacking of a structure for a function other than the one for which it originally evolved is the rule rather than the exception in biology. For example, two of the bones in the lower jaws of reptiles which evolved for chewing became transformed into the tiny middle-ear bones of mammals—used for hearing—simply because these bones were "at the right place at the right time."

I would conjecture that the TPO junction—especially the angular gyri—in the two hemispheres may have also evolved complementary roles in mediating somewhat different types of metaphor: the left one for cross-modal ones (e.g., "loud shirt," "sharp cheese") and the right for spatial metaphors (he "stepped down" from his post). This has not yet been tested systematically. But, I recently tested five patients with left angular gyrus lesions and found that they were abysmal at interpreting proverbs and metaphors.

This failure cannot be attributed to any intellectual deficits or problems with comprehension or expression of language (they had no aphasia other than some difficulty with object naming). Indeed, they often produced elaborate, long-winded, even ingenious interpretations of the proverb, yet completely off the mark, missing the main point (a bit like grant reviewers and referees of scientific journals!).

Finally, I would like to turn to the evolution of language. This has always been a very controversial topic. Language is amazing. Its subtleties and nuances (including what's called recursive embedding) combine with an enormous lexicon to produce a highly sophisticated mechanism. It's easy to imagine a single trait ... like a giraffe's long neck arising from the progressive accumulation of chance variations. But how could an extraordinarily complex mechanism like language with so many interlocking components have evolved through the blind workings of chance—through natural selection? How did we make the transition from the grunts and howls and groans of our ape-like ancestors to all the sophistication of a Shakespeare, or a George W. Bush? There have been several theories. Alfred Russell Wallace said the mechanism is so complicated it couldn't have evolved through natural selection at all and must have resulted from divine intervention. The second theory was by Noam Chomsky, the founding father of linguistics. Chomsky said something quite similar, although he didn't invoke God. He said this mechanism is so sophisticated and elaborate that it couldn't have emerged through natural selection, through the blind workings of chance, but that packing one hundred billion nerve cells in such a tiny space may result in some new laws of physics emerging. He almost says it's a miracle, although he doesn't use the word. Unfortunately, neither Wallace's nor Chomsky's theory can be tested. A third idea was proposed by the brilliant MIT psychologist Steve Pinker. Pinker suggests that the evolution of language is no big mystery. What we see now is the final result of evolution, and it seems mysterious only because we don't know what the intermediate steps were. I suspect he is on the right track—natural selection is the only plausible explanation—but he doesn't go far enough. As biologists, we want the details—the devil's in the details. We want to know what those intermediate steps were, not merely that language *could* have evolved through natural selection. (The technical term for this is "trajectory through the fitness landscape.") And the vital clue to discovering those steps comes from the booba/kiki example, from synesthesia, leading me to propose what I call the synesthetic bootstrapping theory of language origins.

Let's begin with the lexicon. How did we evolve a shared vocabulary—such a huge repertoire of thousands of words?[9] Did our ancestral hominids sit around the fire and say, "Everybody call that object an axe?" Of course not. But if they didn't do that, what did they do? The booba/kiki example provides the clue. It shows there is a pre-existing, *non-arbitrary* translation between the visual appearance of an object represented in the fusiform gyrus and the auditory representation in the auditory cortex. In other words, a synesthetic cross-modal abstraction is already going on, a pre-existing translation, if you like, between visual appearance and auditory representation. Admittedly, this is a very small bias, but that is all that is required in evolution to get something started.[10]

But this is only part of the story. Just as there is a pre-existing, built-in cross-activation between sound and vision—the booba/kiki effect—there's also a non-arbitrary cross-activation between the visual area in the fusiform and the Broca's area in the front of the brain that generates programs which control the muscles of vocalization, phonation and articulation—how we move our lips, tongue and mouth. How do we know that? Say the words "teeny weeny," "un peu," "diminutive." Look at what your lips are doing: they are physically mimicking the visual appearance of what you are saying. Now say "enormous," "large" ... Such mimicry indicates a pre-existing bias to systematically

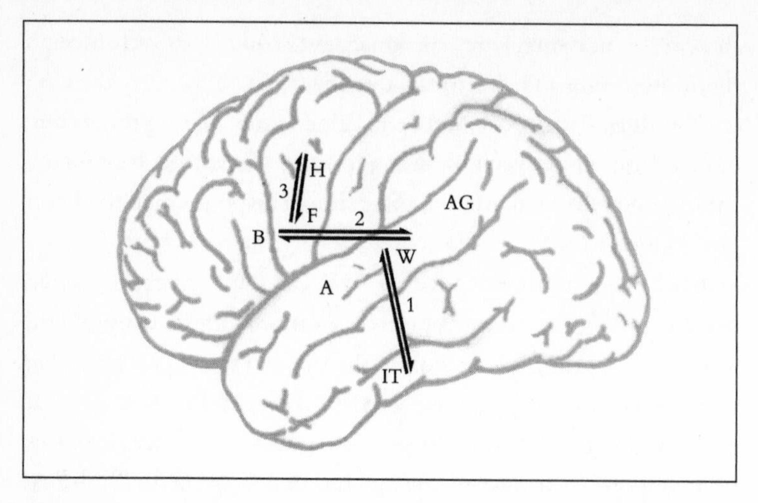

Figure 4.7 *A new synesthetic bootstrapping theory of language origins. Arrows depict cross-domain remapping of the kind we postulate for synesthesia in the fusiform gyrus. (1) A non-arbitrary synesthetic correspondence between visual object shape (as represented in IT and other visual centers) and sound contours represented in the auditory cortex (as in our booba/kiki example). Such synesthetic correspondence could be based on either direct cross-activation or mediated by the angular gyrus – long known to be involved in intersensory transformations. (2) Cross-domain mapping (perhaps involving the arcuate fasciculus) between sound contours and motor maps in or close to Broca's area (mediated, perhaps, by mirror neurons). (3) Motor to motor mappings (synkinesia) caused by links between hand gestures and tongue, lip and mouth movements in the Penfield motor homunculus synkinetically mimic the small pincer gesture made by opposing thumb and index finger (as opposed to "large" or "enormous").*

The role of the angular gyrus (with an excess of cross-connections) that we have postulated for mediating higher synesthesia can be tested directly by silencing it temporarily using transcranial magnetic stimulation in volunteers.

map certain visual shapes on to certain "sounds" represented in the motor maps in the Broca's area (Figure 4.7).

The third part of my theory is that there is also a pre-existing cross-activation between the hand area and the mouth area, which are right next to each other in the Penfield motor map in the brain (see Figure 1.6). I can illustrate with an example from Charles Darwin. He noticed that when people cut something with a pair of scissors they clench and unclench their jaws unconsciously, as if to echo or mimic the movements of the fingers. I call this phenomenon synkinesia, because the hand and mouth areas are right next to each other and perhaps there is some spill-over of signals from gestures to vocalizations (e.g., the oral gestures for "little" or "diminutive" or "teeny weeny") synkinetically mimic the small pincer gesture of opposing the thumb and index finger.

A system of non-verbal communication would have been important to our ancestral hominids unable to engage in loud vocalization when hunting. We also know that the right hemisphere produces guttural emotional utterances along with the anterior cingulate. Combine these with the pre-existing translation of manual gesturing into mouth, lip and tongue movements and you get humankind's first words: proto-words.

Thus we have three things in place—first, hand to mouth; second, mouth in Broca's area to visual appearance in the fusiform and sound contours in the auditory cortex; and third, auditory to visual, the booba/kiki effect. Acting together, these three have a synergistic bootstrapping effect—an avalanche culminating in the emergence of a primitive language (Figure 4.7).

All this is fine, but how do we explain the hierarchical structure of syntax? For example, "He knows that I know that he knows that I had an affair with his wife," or, "She hit the boy who kissed the girl that she disliked." How does this hierarchic embedding in language come about? Partly, I think, from semantics, from the region of the TPO junction which is involved in abstraction (I have already explained how abstraction might have evolved). It is possible that abstraction (and semantics) feeds into syntactic structure and may have played a role in "guiding" its evolution. But partly, the hierarchical "tree" structure of syntax may have evolved from tool use. Early hominids were very good at tool use, especially what is known as the sub-assembly technique: a piece of flint is made into a head—step one—then hafted on to a handle—step two—then the whole thing is used as a tool or weapon—step three. There is a close operational analogy between this function and the embedding of noun clauses within longer sentences. So perhaps what originally evolved for tool use in the hand area is now exapted and assimilated in the Broca's area to be used in aspects of syntax such as hierarchic embedding.

Each of these effects is a small bias, but acting in conjunction they may have paved the way for the emergence of sophisticated language. This is very different from Steve Pinker's idea that language is a specific adaptation which evolved step by step for the sole purpose of communication. I suggest, instead, that it is the fortuitous synergistic combination of a number of mechanisms which evolved for other purposes initially that later became assimilated into the mechanism that we call language. This often happens in evolution, but this style of thinking has yet to permeate neurology and psychology. It seems odd that neurologists often overlook evolution as explanation, given that, as Dobzhansky once said, nothing in biology makes sense except in the light of evolution.[11]

My final point concerns the mirror neurons mentioned in chapter 2—cells in the parietal and frontal lobes that fire not only when you move your hand but also when you watch someone else move their hand. Similar neurons also exist for orofacial movements: not only do they fire when you stick your tongue out or purse your lips, but also when you see someone else do it, even though you have never directly seen your own lips or tongue. The neurons must be establishing a congruence between (1) the highly specific volitional motor command sequence sent to muscles of phonation and articulation; (2) *felt* lip and tongue position (proprioception) from sensors in your oral muscles; (3) the *seen* image of someone else's lips and tongue; and (4) the *heard* phoneme. This ability is partly innate—if you stick your tongue out at a newborn baby, it will imitate you—but the more complex three-way congruence between the phoneme sound, the appearance of lips and tongue and their felt position that is required for lipreading is probably acquired in childhood. It seems likely that these neurons may have played an important role in developing a shared vocabulary by miming seen vocalizations and establishing a congruence with heard sounds.

As an illustration, consider the fact that if a normal English-speaking person watches me silently pronounce the syllable Rrrrr or Llll he can "lipread" and correctly infer which one I am producing, presumably using his mirror neurons. But when I recently tried this on a native Chinese speaker (who learned English only as an adult) he had considerable difficulty making this judgment; perhaps the mirror neurons required for making this particular distinction simply didn't develop in him.

In this chapter, we began with a disorder—synesthesia—that's been known for a century but treated mainly as a curiosity. We then showed that the phenomenon is real, pointed out what the underlying brain mechanisms might be, and spelled out its broader implications. (One day we might be able to clone the gene or genes, if we can find a large enough family. I have recently heard rumors of a whole island of synesthetes!) We then suggested that if the gene is expressed in the fusiform gyrus it results in lower synesthesia, and higher synesthesia if expressed in the angular gyrus. If it's expressed all over you get artsy types! We have learned something of the detailed perceptual psychophysics—such as the pop-out effect of 2s against 5s, which can be measured—and can perhaps begin to approach elusive phenomena such as abstract thought, metaphor, Shakespeare, even the evolution of language. All this by studying one little quirk that people call synesthesia. So I agree wholeheartedly with what Thomas Henry Huxley said in the nineteenth century: contrary to the views of Bishop Wilberforce and Benjamin Disraeli, we are not angels, we are merely sophisticated apes. Yet we don't *feel* like that—we feel like angels trapped inside the bodies of beasts, forever craving transcendence, trying to spread our wings and fly off. Really a very odd predicament to be in, if you think about it.

Our revels now are ended. These our actors,
As I foretold you, were all spirits and
Are melted into air, into thin air …
We are such stuff
As dreams are made on,
And our little life
Is rounded with a sleep.

5

Neuroscience — The New Philosophy

"All of philosophy consists of unlocking, exhuming, and recanting what's been said before, and then getting riled up about it."

V. S. RAMACHANDRAN

The main theme of the book so far has been the idea that the study of patients with neurological disorders has implications far beyond the confines of medical neurology, implications even for the humanities, for philosophy, maybe even for aesthetics and art. In this final chapter I would like to continue this theme and take up the challenge of mental illness. The boundary between neurology and psychiatry is becoming increasingly blurred and it is only a matter of time before psychiatry becomes just another branch of neurology. I shall also touch on a few philosophical issues such as free will and the nature of self.

There have traditionally been two different approaches to mental illness. The first one tries to identify chemical imbalances, changes in transmitters and receptors in the brain, and attempts to correct these changes using drugs. This approach has revolutionized psychiatry and has been phenomenally successful. Patients who used to be put in straitjackets or locked up can now lead relatively normal lives. The second approach we can loosely characterize as the so-called Freudian approach. It assumes that most mental illness arises from early upbringing. I'd like to propose a third approach which is radically different from either of these but which, in a sense, complements them both.

To understand the origins of mental illness it is not enough merely to say that some transmitter has changed in the brain. We need to know how that change produces the bizarre symptoms that it does—why certain patients have certain symptoms and why those symptoms are different for different types of mental illness. I will attempt to explain the symptoms of mental illness in terms of what is known about function, anatomy and neural structures in the brain. And I will suggest that many of these symptoms and disorders seem less bizarre when viewed from an evolutionary standpoint, that is from a Darwinian perspective. I also propose to give this discipline a new name—evolutionary neuro-psychiatry.

Let us begin with the classic example of what most people consider to be a purely mental disorder or psychological disturbance—hysteria. "Hysteria" is used here in its strict medical sense, as opposed to the everyday notion of a person shouting and screaming. In the strictly medical sense, a hysteric is a patient who suddenly experiences blindness or develops a paralysis of an arm or a leg, but who has no neurological deficits that could be responsible for his or her condition: a brain MR scan reveals that the brain is apparently completely normal, there are no identifiable lesions, no apparent damage. So the symptoms are dismissed as being purely psychological in origin.

However, recent brain-imaging studies using PET scans and functional Magnetic Resonance imaging have dramatically changed our understanding of hysteria. Using PET scans and fMR, we can now find what parts of the brain are active or inactive when a patient performs a specific action or engages in a specific mental process. For example, when we do mental arithmetic, the left angular gyrus usually exhibits activity. Or if I were to prick you with a needle and cause pain, another part of your brain would light up. We can then conclude that the particular brain region that lights up is somehow involved in mediating that function.

If you were to wiggle your finger, a PET scan would reveal that two areas of your brain light up. One is called the motor cortex, which is actually sending messages to execute the appropriate sequence of muscle twitches to move your finger, but there is another area in front of it called the pre-motor cortex that *prepares* you to move your finger.

John Marshall, Chris Frith, Richard Frackowiak, Peter Halligan and others tried this experiment on a hysterically paralyzed patient. When he tried to move his leg, the motor area failed to light up even though he claimed to be genuinely intending to move his leg. The reason he was unable to is that another area was simultaneously lighting up: the anterior cingulate and the orbito-frontal lobes. It's as if this activity in the anterior cingulate and orbito-frontal cortex was inhibiting or vetoing the hysterical patient's attempt to move his leg. This makes sense, because the anterior cingulate and orbito-frontal cortex are intimately linked to the limbic emotional centers in the brain, and we know that hysteria originates from some emotional trauma that is somehow preventing him from moving his "paralyzed" leg.

Of course, all this doesn't explain exactly why hysteria occurs, but now we at least know where to look. In the future it might be possible to use a brain scan to distinguish genuine hysterics from malingerers or fraudulent insurance claimants. And it does prove that one of the oldest "psychological" disturbances—one that Freud studied—has a specific and identifiable organic cause. (Actually, an important control is missing in this experiment: no one has yet obtained a brain scan from a genuine malingerer.)

We can think of hysteria as a disorder of "free will," and free will is a topic that both psychologists and philosophers have been preoccupied with for over two thousand years.

Several decades ago the American neurosurgeon Benjamin Libet and the German physiologist Hans Kornhuber were experimenting on volunteers exercising free will, instructing subjects to, for example, wiggle a finger at any time of their own choosing within a ten-minute period. A full three-quarters of a second *before* the finger movement the researchers picked up a scalp EEG potential, which they called the "readiness potential," even though the subject's sensation of consciously willing the action coincided almost exactly with the actual onset of finger movement. This discovery caused a flurry of excitement among philosophers interested in free will. For it seemed to imply that the brain events monitored by the EEG kick in almost a second before there is any sensation of "willing" the finger movement, even though your *subjective* experience is that your will caused the finger movement! But how can your will be the cause if the brain commands begin a second earlier? It's almost as though your brain is really in charge and your "free will" is just a post-hoc rationalization—a delusion, almost—like King Canute thinking he could control the tides or an American president believing that he is in charge of the whole world.

This alone is strange enough, but what if we add another twist to the experiment. Imagine I'm monitoring your EEG while you wiggle your finger. Just as Kornhuber and Libet did, I will see a readiness potential a second before you act. But suppose I display the signal on a screen in front of you so that you can *see* your free will. Every time you are about to wiggle your finger, supposedly using your own free will, the machine will tell you a second in advance! What would you now experience? There are three logical possibilities. (1) You might experience a sudden loss of will, feeling that the machine is controlling you, that you are a mere puppet and that free will is just an illusion. You may even become paranoid as a result, like schizophrenics who think their actions are controlled by aliens or implants (I'll return to this later). (2) You might think that it does not change your sense of free will one iota, preferring to believe that the machine has some sort of spooky paranormal precognition by which it is able to predict your movements accurately. (3) You might confabulate, or rearrange the experienced sequence mentally in order to cling to your sense of freedom; you might deny the evidence of your eyes and maintain that your sensation of will preceded the machine's signal, not vice versa.

At this stage this is still a "thought experiment"—technically it is hard to get a feedback EEG signal on each trial, but we are trying to get around this obstacle. Nevertheless, it is important to note that one can do experiments that have direct relevance to broad philosophical issues such as free will—a field in which my colleagues Pat Churchland, Dan Wegner and Dan Dennett have all made valuable contributions.

Leaving aside this "thought experiment" for the moment, let's return to the original observation on the readiness potential with its curious implication that the brain events are kicking in a second or so before any actual finger movement, even though conscious intent to move the finger coincides almost exactly with the wiggle. Why might this be happening? What might the evolutionary rationale be?

The answer is, I think, that there is an inevitable neural delay before the signal arising in one part of the brain makes its way through the rest of the brain to deliver the message: "wiggle your finger." (A televisual equivalent is the sound delay experienced when conducting an interview via satellite.) Natural selection has ensured that the subjective sensation of willing is delayed deliberately to coincide not with the onset of the brain commands but with the actual execution of the command by your finger.[1]

And this in turn is important because it means that the subjective sensations which accompany brain events must have an evolutionary purpose. For if that were not the case, if they merely *accompanied* brain events, as so many philosophers believe (this is called epiphenomenalism)—in other words, if the subjective sensation of willing is like a shadow that accompanies us as we move but is not causal in making us move—then why would evolution bother delaying the signal so that it coincides with our movement?

So we have a paradox: on the one hand, the experiment shows that free will is illusory: it cannot be causing the brain events because the events kick in a second earlier. But on the other hand, the delay must have some function, otherwise why would the delay have evolved? Yet if it *does* have a function, what could it be other than moving (in this case) the finger? Perhaps our very notion of causation requires a radical revision ... as happened in quantum mechanics.

Other types of "mental" illness can also be approached, perhaps, through brain imaging. Take the case of pain: when someone is jabbed with a needle, there is usually activity in many regions of the brain, but especially in the insula and in the anterior cingulate. The former structure seems to be involved in sensing the pain and the latter in giving pain its aversive quality. So when the pathways leading from the insula to the anterior cingulate are severed, the patient can feel the pain, but it doesn't hurt—a paradoxical syndrome called pain asymbolia. This leads me to wonder about the image of the brain of a masochist who derives pleasure from pain, or a patient with Lesch–Nyhan syndrome who "enjoys" mutilating himself. The insula would be activated, of course, but would the anterior cingulate also light up? Or, given, especially, the sexual overtones of masochism, a region concerned with pleasure, such as the nucleus accumbens, septum or hypothalamic nuclei? At what stage in processing do the "pain/pleasure" labels get switched? (I am reminded of the masochist from Ipswich who loved taking ice cold showers at four in the morning and therefore didn't.)

In chapter 1 I mentioned the Capgras delusion, sometimes seen in patients who have sustained a head injury, in which sufferers start claiming that someone they both recognize and know well—such as their mother—is an imposter.

The theoretical explanation for Capgras syndrome is that the connection between the visual areas and the emotional core of the brain, the limbic system and the amygdala, has been cut by the accident (Figure 1.3). So when the patient looks at his mother, since the visual areas in the brain concerned with recognizing faces is not damaged, he is able to say that she *looks* like his mother. But there is no emotion, because the wire taking that information to the emotional centers is cut, so he tries to rationalize this by believing her to be an imposter.

How can this theory be tested? Well, it is possible to measure the gut-level emotional reaction that someone has to a visual stimulus—or any stimulus—by measuring the extent to which they sweat. When any of us sees something exciting, emotionally important, the neural activation cascades from the visual centers to the emotional centers in the brain and we begin to sweat in order to dissipate the heat that we are going to generate from exercise, from action (feeding, fleeing, fighting, or sex). This effect can be measured by placing two electrodes on a person's skin to track changes in skin resistance—when skin resistance falls, we call it a galvanic skin response. Familiar or nonthreatening objects or people produce no galvanic skin response because they generate no emotional arousal. But if you look at a lion or a tiger, or—as it turns out—your mother, a huge galvanic skin response occurs. Believe it or not, every time you see your mother, you sweat! (And you don't even have to be Jewish.)

But we found that this doesn't happen in Capgras patients, supporting the idea that there has been a disconnection between vision and emotion.

There exists an even more bizarre disorder, Cotard's syndrome, in which the patient starts claiming he or she is dead. I suggest that this is similar to Capgras except that instead of vision alone being disconnected from the emotional centers in the brain, all the senses become disconnected from the emotional centers. So that nothing in the world has any emotional significance, no object or person, no tactile sensation, no sound—nothing—has emotional impact. The only way in which a patient can interpret this complete emotional desolation is to believe that he or she is dead. However bizarre, this is the only interpretation that makes sense to him; the reasoning gets distorted to accommodate the emotions. If this idea is correct we would expect no galvanic responses in a Cotard's patient whatever the stimulus.

The delusion of Cotard's is notoriously resistant to intellectual correction. For example, a man will agree that dead people cannot bleed; then, if pricked with a needle, he will express amazement and conclude that the dead *do* bleed after all, instead of giving up his delusion and inferring that he is alive. Once a delusional fixation develops, all contrary evidence is warped to accommodate it. Emotion seems to override reason rather than the other way around. (Of course, this is true of most of us to some extent. I have known many an otherwise rational and intelligent person who believes the number 13 to be unlucky or who won't walk under a ladder.)

Capgras and Cotard's are both rare syndromes, but there is another disorder, a sort of mini-Cotard's, that is much more commonly seen in clinical practice. This disorder is known as derealization and depersonalization, and is found in acute anxiety, panic attacks, depression and other dissociative states. Suddenly the world seems completely unreal—like a dream. The patient feels like a zombie.

I believe such feelings involve the same circuitry as Capgras and Cotard's. In nature, an opossum when chased by a predator will suddenly lose all muscle tone and play dead, hence the phrase "playing possum." This is a good strategy for the opossum because (a) any movement will encourage the predatory behavior of the carnivore and (b) carnivores usually avoid carrion, which might be infected. Following the lead of Martin Roth, Mauricio Sierra and German Berrios, I suggest that derealization and depersonalization, and other dissociative states, are examples of playing possum in the emotional realm and that this is an evolutionary adaptive mechanism.

There is a well-known story of the explorer David Livingstone being attacked by a lion. He saw his arm being mauled but felt no pain or even fear. He felt detached from it all, as if he were watching events from a distance. The same thing can happen to soldiers in battle or to a woman being raped. During such dire emergencies, the anterior cingulate in the brain, part of the frontal lobes, becomes extremely active. This inhibits or temporarily shuts down the amygdala and other limbic emotional centers, so temporarily suppressing potentially disabling emotions such as anxiety and fear. But at the same time, the anterior cingulate activation generates extreme alertness and vigilance in preparation for any appropriate defensive reaction that might be required.

In an emergency, this James Bond-like combination of shutting down emotions ("nerves of steel") while being hypervigilant is useful, keeping us out of harm's way. It is better to do nothing than to engage in some sort of erratic behavior. But what if the same mechanism is accidentally triggered by chemical imbalances or brain disease, when there is no emergency? A person looks at the world, is intensely alert, hypervigilant, but the world has become completely devoid of emotional meaning because the limbic system has been shut down. There are only two possible ways to interpret this strange predicament, this paradoxical state of mind. Either "the world isn't real"—derealization—or "I am not real"—depersonalization.

Epileptic seizures originating in this part of the brain can also produce these dreamlike states of derealization and depersonalization. Intriguingly, we know that during a seizure, when the patient is experiencing derealization, there is no galvanic skin response to anything. Following the seizure, skin response returns to normal. All of which supports the hypothesis that we have been considering.

Probably the disorder most commonly associated with the word "madness" is schizophrenia. Schizophrenics do indeed exhibit bizarre symptoms. They hallucinate, often hearing voices. They become delusional, thinking they're Napoleon or Ramachandran. They are convinced the government has planted devices in their brain to monitor their thoughts and actions. Or that aliens are controlling them.

Psycho-pharmacology has revolutionized our ability to treat schizophrenia, but the question remains: why do schizophrenics behave as they do? I'd like to speculate on this, based on some work my colleagues and I have done on anosognosia (denial of illness)—which results from right-hemisphere lesions—and some very clever speculations by Chris Frith, Sarah Blakemore and Tim Crow. Their idea is that, unlike normal people, schizophrenics cannot tell the difference between their own internally generated images and thoughts and perceptions that are evoked by real things outside.

If I conjure up a mental picture of a clown in front of me, I don't confuse it with reality partly because my brain has access to the internal command I gave. I am expecting to visualize a clown, and that is what I see. It is not an hallucination. But if the "expectation" mechanism in my brain that does this becomes faulty, then I would be unable to tell the difference between a clown I'm imagining and a clown I'm actually seeing there. In other words, I would believe that the clown was real. I would hallucinate, and be unable to differentiate between fantasy and reality.

Similarly, I might momentarily entertain the thought that it would be nice to be Napoleon, but in a schizophrenic this momentary thought becomes a full-blown delusion instead of being vetoed by reality.

What about the other symptoms of schizophrenia—alien control, for example? A normal person knows that he moves of his or her own free will, and can attribute the movement to the fact that the brain has sent the command "move." If the mechanism that monitors intention and compares it with performance is flawed, a more bizarre interpretation is likely to result, such as that body movements are controlled by aliens or brain implants, which is what paranoid schizophrenics claim.

How do you test a theory like this? Here is an experiment for you to try: using your right index finger, tap repeatedly your left index finger, keeping your left index finger steady and inactive.

Notice how you feel the tapping mainly on the left finger, very little on the right finger. That is because the brain has sent a command from the left hemisphere to the right hand saying "move." It has alerted the sensory areas of the brain to expect some touch signals on the right hand. Your left hand, however, is perfectly steady, so the taps upon it come as something of a surprise. This is why you feel more sensation in the immobile finger, even though the tactile input to both fingers is exactly the same. (If you change hands, you will find that the results are reversed.)

Following our theory, I predict that if a schizophrenic were to try this experiment, he would feel the sensations equally in both fingers since he is unable to differentiate between internally generated actions and externally generated sensory stimuli. It's a five-minute experiment—yet no one has ever tried it.[2]

Or imagine that you are visualizing a banana on a blank white screen in front of you. While you are doing this, if I secretly project a very low-contrast physical image of the banana on the screen, your threshold for detecting this real banana will be elevated—presumably even your normal brain tends to get confused between a very dim real banana and one which you imagine. This surprising result is called the "Perky effect" and one would predict that it would be amplified enormously in schizophrenics.

Another simple yet untried experiment: as you know, you can't tickle yourself. That is because your brain knows you're sending the command. Prediction: a schizophrenic will laugh when he tickles himself.

Even though the behavior of many patients with mental illness seems bizarre, we can now begin to make sense of the symptoms using our knowledge of basic brain mechanisms. Mental illness might be thought of as disturbances of consciousness and of self, two words that conceal great depths of ignorance. Let me try to summarize my own view of consciousness. There are really two problems here—the problem of the subjective sensations or qualia and the problem of the self. The problem of qualia is the more difficult.

The qualia question is, how does the flux of ions in little bits of jelly—the neurons—in our brains give rise to the redness of red, the flavor of Marmite or paneer tikka masala or wine?[3] Matter and mind seem so utterly unlike each other. One way out of this dilemma is to think of them really as two different ways of describing the world, each of which is complete in itself. Just as we can describe light as made up either of particles or as waves—and there's no point in asking which description is correct, because they both are, even though the two seem utterly dissimilar—the same may be true of mental and physical events in the brain.

But what about the self, the last remaining great mystery in science and something that everybody is interested in? Obviously self and qualia are two sides of the same coin. You can't have free-floating sensations or qualia with no one to experience them and you can't have a self completely devoid of sensory experiences, memories or emotions. (As we saw in Cotard's syndrome, when sensations and perceptions lose all their emotional significance and meaning, the result is a dissolution of self.)

What exactly is meant by the "self"? Its defining characteristics are fivefold. First of all, continuity: a sense of an unbroken thread running through the whole fabric of our experience with the accompanying feeling of past, present and future. Second, and closely related, is the idea of unity or coherence of self. In spite of the diversity of sensory experiences, memories, beliefs and thoughts, we each experience ourselves as one person, as a unity.

Third is a sense of embodiment or ownership—we feel ourselves anchored to our bodies. Fourth, a sense of agency, what we call free will, being in charge of our own actions and destinies. I can wiggle my finger but I can't wiggle my nose or your finger.

Fifth, and most elusive of all, the self, almost by its very nature, is capable of reflection—of being aware of itself. A self that's unaware of itself is an oxymoron.

Any or all of these different aspects of self can be differentially disturbed in brain disease, which leads me to believe that the self comprises not just one thing, but many. Like "love" or "happiness," we use one word, "self," to lump together many different phenomena. For example, if I stimulate your right parietal cortex with an electrode (while you're conscious and awake), you will momentarily feel that you are floating near the ceiling, watching your own body down below. You have an out-of-the-body experience. The embodiment of self—one of the axiomatic foundations of your self—is temporarily abandoned.[4] And this is true of all of those aspects of self I listed above. Each of them can be selectively affected in brain disease.

Keeping this in mind, I see three ways in which the problem of self might be tackled by neuroscience. First, maybe the problem of self is a straightforward empirical one. Maybe there is a single, very elegant, Archimedes-type Eureka! solution to the problem, just as DNA base-pairing was the solution to the riddle of heredity. I think this is unlikely, but I could be wrong.

Second, given my earlier remarks about the self, the notion of the self as being defined by a set of attributes—embodiment, agency, unity, continuity—maybe we will succeed in explaining each of these attributes individually in terms of what is going on in the brain. Then the problem of what the self is will vanish, or at least recede into the background, just as scientists no longer speak of vital spirits or ask what "life" is. (We recognize that life is a word loosely applied to a collection of processes—DNA replication and transcription, Krebs cycle, Lactic acid cycle, etc., etc.)

Third, maybe the solution to the problem of the self is not a straightforward empirical one. It may instead require a radical shift in perspective, the sort of thing that Einstein did when he rejected the assumption that things can move at arbitrarily high velocities. When we finally achieve such a shift in perspective, we may be in for a big surprise and find that the answer was staring at us all along. I don't want to sound like a New Age guru, but there are curious parallels between this idea and the Hindu philosophical (albeit somewhat nebulous) view that there is no essential difference between self and others, or that the self is an illusion.

Of course, I have no clue what the solution to the problem of self is, what the shift in perspective might be. If I did I would dash off a paper to *Nature* today, and overnight I'd become the most famous scientist alive. But, just for fun, I'll have a crack at describing what the solution might look like.

I will begin with qualia. It seems quite obvious that qualia must have evolved to fullfil a specific biological function—they cannot be mere by-products (an "epiphenomenon") of neural activity. In 1997 I suggested that sensory representations that are themselves devoid of qualia might acquire qualia in the process of being economically encoded or "prepared" into manageable chunks as they are delivered to a central executive structure higher up in the brain. The result is a higher order representation that serves new computational goals. Let us call this second, higher-order, representation a metarepresentation. (Though I feel a bit uncomfortable using the prefix "meta," which is often employed as a disguise for fuzzy thinking—especially among social scientists.) One could think of this metarepresentation almost as a second "parasitic" brain—or at least a set of processes—that has evolved in us humans to create a more economical description of the rather more automatic processes that are being carried out in the first brain. Ironically this idea implies that the so-called homunculus fallacy—the notion of a "little man in the brain watching a movie screen filled with qualia"—isn't really a fallacy. In fact, what I am calling a metarepresentation bears an uncanny resemblance to the homunculus that philosophers take so much delight in debunking. I suggest that the homunculus is simply either the metarepresentation itself, or another brain structure that emerged later in evolution for creating metarepresentations, and that it is either unique to us humans or considerably more sophisticated than a "chimpunculus." (Bear in mind, though, that it doesn't have to be a single new structure—it could be a set of novel functions that involves a distributed network. Ideas similar to this have also been foreshadowed by David Darling, Derek Bickerton, Marvin Minsky and many others, although usually invoked for reasons other than the ones I consider here.)

But what is the purpose of creating such a metarepresentation? Clearly it cannot be just a copy or duplicate of the first—that would serve no purpose. Just like the first representation itself, the second one serves to emphasize or highlight certain aspects of the first in order to create tokens that facilitate novel styles of subsequent computation, either for internally juggling symbols sequentially ("thought") or for communicating ideas to others through a one-dimensional sound stream ("language"). Indeed, if you combine abstraction (discussed at length in chapter 4) with sequential symbol juggling you get "thinking"—a hallmark of our species.

Once this line was crossed in evolution the brain became capable of generating what Karl Popper would call "conjectures"; it could tentatively try out novel—even absurd—juxtapositions of perceptual tokens just to see what would happen. It's a moot point whether an ape can conjure up a visual image of a horse it has just seen, but it is unlikely that it can visualize a horse with a horn—a unicorn—or imagine a cow with wings—something humans can do effortlessly.

These ideas lead to an interesting question. Are qualia and self-awareness unique to humans or are they present in other primates? And to what extent do they depend on language? Vervet monkeys in the wild have specific calls to warn their companions about different predators. A "tree snake" call will send them scurrying down to the ground and a "terrestrial leopard" call will send them climbing higher up the tree. But the caller doesn't *know* that it is warning the others; vervet monkeys have no introspective consciousness—which, as we have seen, probably requires another part of the brain (perhaps linked to aspects of language) to generate a representation of the earlier sensory representation (a metarepresentation) of the snake or leopard. We could teach a monkey that a pig is dangerous by administering a mild electric shock whenever the pig appears. But what if that monkey were put back in the treetops and a pig lifted on to an adjoining branch? I predict that the monkey would become agitated but not be capable of generating the "snake" cry to warn the other monkeys to climb down; i.e., to start using it as a verb. It is very likely that only humans are capable of the kind of consciousness of one's qualia and of the limits on one's capabilities—"will power"—that this would require.

Which parts of the brain are involved in these novel styles of computation? A tentative list would include the amygdala (that gauges emotional significance), structures like the angular gyrus and Wernicke's area that are clustered around the left temporo-parieto-occipital (TPO) junction, and the anterior cingulate, involved in "intention." As I noted in *Phantoms in the Brain*, another reason for choosing the temporal lobes—especially the left temporal lobe ... is that this is where much of language—especially semantics—is represented. If I see an apple, it is the activity in the temporal lobes that allows me to apprehend all its implications almost simultaneously. Recognition of it as a fruit of a certain type occurs in IT (inferotemporal cortex), the amygdala gauges its significance for my well-being, and Wernicke's and other areas alert me to all the subtle nuances of meaning that the mental image including the word apple evokes; I can eat the apple, I can smell it, I can bake a pie, remove its pith, plant its seeds, use it to "keep the doctor away," tempt Eve, and on and on. The choice of which implication occupies center stage ("attention") and what to do about it is partly mediated by the anterior cingulate. When it becomes damaged the patient appears fully awake but loses all desire to talk, think, choose or act; he suffers from "akinetic mutism."

An important question that emerges from all this is the extent to which the metarepresentation is linked to the emergence of language comprehension/meaning capacity.[5] One way to find out would be to see if a patient with Wernicke's aphasia—caused by damage to the language area in the left hemisphere—can lie non-verbally, even though he is incapable of understanding or engaging in meaningful conversation. For unless you have an explicit representation of your representations you cannot begin to distort them before transmitting them to others; i.e., you cannot lie.[6] (This is because if the first representation itself is distorted you deceive yourself—which defeats the whole purpose of lying. It might help you disseminate your genes to lie to a potential mate that you have a huge bank balance, but if you actually believe it—if you are delusional—you might start spending money you don't have.)

Indeed, deliberate lying is the litmus test of whether a subject—be it a chimp or an infant or a brain-damaged individual—is simultaneously capable of both modeling other people's minds *and* of reflective self-consciousness. It is true that a bird can feign a broken wing to distract a predator away from her chicks, but she doesn't realize she's doing this; she doesn't have a representation of the representation. And therefore she cannot deploy this strategy in an open-ended manner in novel situations where it might prove useful. For example, she cannot feign this injury to invite more attention and compassion from her mate (though such an ability could evolve later through natural selection).

The distinction between deliberate lying and self-deception becomes very blurred in disorders like anosognosia (chapter 2) in which a patient with a paralyzed left arm caused by right hemisphere damage will deny her paralysis. Oddly enough, when I asked one of these patients whether she could touch my nose with her left hand she said, "Sure … but watch out—I might poke your eye!" And on another occasion when I asked a retired Army general "Can you use your left arm?" his reply was "Yes—but I won't. I am not accustomed to taking orders, doctor." Such remarks imply that somebody in there "knows" the truth and it leaks out even though "she" or "he"—the reflectively conscious person—does not. ("Methinks the lady doth protest too much." Shades of Freudian psychology here.) And, again, I would point out that the very existence of the phenomenon of self-deception implies that there must be a self to deceive. Far from being an epiphenomenon, the sense of self must have evolved through natural selection to enhance survival and, indeed, must include within it the ability to preserve its integrity and stability—even deceiving itself when necessary. I doubt very much that an ape would be capable of Freudian defense mechanisms such as a "nervous laugh," a denial or a rationalization (assuming we could even test it through signing).

All this takes us back to my earlier remark that qualia and self are really two sides of a coin—you can't have one without the other. The ability to use special brain circuits to create metarepresentations[7] of sensory and motor representations—partly to facilitate language and partly facilitated *by* language—might have been critical for the evolution of both full-fledged qualia and a sense of self. As we noted earlier, it is impossible to have free-floating qualia without a self experiencing it, nor a self existing in isolation, devoid of all feeling and sensation.

A similar distinction can be made between representations of "raw" emotions and metarepresentations of them that would allow you to reflect on the emotion and make sophisticated choices—even withholding certain actions that might otherwise follow automatically. If I sprinkle pepper near your nose you sneeze reflexively—but why is this accompanied by a distinctive sneeze quale? (Unlike the knee-jerk, which can occur without quale even in a paraplegic.) Ironically this quale may have evolved as a metarepresentation for the sole purpose of enabling you to abort the sneeze voluntarily if you need to (e.g., when pursuing game). A cat probably cannot voluntarily abort an impending sneeze, given that it doesn't—in my view—have a metarepresentation. A sneeze can hardly be described as an emotion, but the same principle probably applies to more complex human emotions. A cat simply pounces when it sees a long black moving shape but cannot contemplate a mouse or "mousiness" the way you and I do. Nor can it experience subtle emotions like humility, arrogance, mercy, desire (as opposed to need) or "tears of self-pity," all of which are based on metarepresentations of emotions requiring complex interactions with social values represented in the orbitorontal cortex. Although emotions are phylogenetically ancient and often regarded as more primitive, in humans they are probably just as sophisticated as reason.

The sense of "unity" of self also deserves comment. Why do you feel like "one" despite being immersed in a constant flux of sensory impressions, thoughts and emotions? This is a tricky question and may well turn out to be a pseudo-problem. Perhaps the self by its very nature can be experienced only as a unity. Actually experiencing two selves may be logically impossible, because it would raise the question of who or what is experiencing the two selves. True, we sometimes speak of "being in two minds," but that is nothing more than a figure of speech. Even people with so-called multiple personality disorder don't experience two personalities simultaneously – the personalities tend to rotate and are mutually amnesic: at any given instant the self occupying center stage is walled off from (or only dimly aware of) the other(s). Even in the extreme case of a split-brain patient whose two hemispheres have been surgically disconnected,[8] the patient doesn't experience doubling subjectively; each hemisphere's "self" is aware only of itself – although it may intellectually deduce the presence of the other.[9]

Another "paradox" of sorts is that even though the self is private – almost by definition – it is very much enriched by social interactions and, indeed, may have evolved mainly in a social context, as both Nick Humphrey and Horace Barlow first pointed out in a conference that Brian Josephson and I organized in 1979.

Let me expand on this. Our brains are essentially model-making machines. We need to construct useful, virtual reality simulations of the world that we can act on. Within the simulation, we need also to construct models of other people's minds because we primates are intensely social creatures. (This is called "a theory of other minds.") We need to do this so that we can predict their behavior. For example, you need to know whether another's action in jabbing you with an umbrella was willful, and so likely to be repeated, or involuntary, in which case it's quite benign. Furthermore, for this internal simulation to be complete it needs to contain not only models of other people's minds but also a model of itself, of its own stable attributes, its personality traits and the limits of its abilities – what it can and cannot do. It is possible that one of these two modeling capacities evolved first and then set the stage for the other. Or—as often happens in evolution—the two may have co-evolved and enriched each other enormously, culminating in the reflective self-awareness that characterizes *Homo sapiens*.

At a very rudimentary level we are reminded of this reciprocity of "self" and "others" each time a newborn baby mimics an adult's behavior. Stick your tongue out at a newborn baby and the baby will stick its tongue out too, poignantly dissolving the boundary, the arbitrary barrier, between self and others. To do this it must create an internal model of your action and then re-enact it in its own brain. An astonishing ability, given that it cannot even see its own tongue, and so must match the visual appearance of your tongue with the felt position of its own. We now know that this is carried out by a specific group of neurons, in the frontal lobes, called the mirror neurons. I suspect that these neurons are at least partly involved in generating our sense of "embodied" self-awareness as well as our "empathy" for others. No wonder children with autism—who (I conjecture) have a deficient mirror-neuron system—are incapable of constructing a theory of other minds, lack empathy, and also engage in self-stimulation to enhance their sense of being a self anchored in a body. It would be interesting to see if an autistic baby (diagnosed sufficiently young) would mimic the tongue protrusion of an adult in the same way that normal infants do.

Without a "theory of other minds" an organism (or person) would also be incapable of blushing—the external marker of embarrassment. (As someone said: "Only humans blush—or need to.") Blushing is a fascinating topic that greatly intrigued

Darwin. Since it is an involuntary "flag" of violation of a social taboo, it may have evolved in humans as a "marker" or index of reliability. When courting a man, a blushing woman is saying (in effect): "I can't lie to you about my affair or cuckold you without my blush giving me away – I'm reliable, so come disseminate your genes through me." If this is correct one might expect that autistic children are incapable of blushing.

In addition to their obvious role in empathy, "mind reading" and evolution of language (chapter 4), mirror neurons may have also played a vital role in the emergence of another important capacity of our minds – namely, learning through imitation – and therefore the transmission of culture.[10] Polar bears had to go through millions of years of natural selection of genes to evolve a fur coat, but a human child can acquire the skill required to make a coat by simply watching his parent slaying a bear and skinning it. Once the mirror neuron system became sophisticated enough, this remarkable ability – imitation and mimesis – liberated humans from the constraints of a strictly gene-based evolution, allowing them to make a rapid transition to Lamarckian evolution. As noted in chapter 2, the result was a rapid horizontal spread and vertical transmission of cultural innovations of the kind that took place about 50,000 to 75,000 years ago leading to the so-called great leap forward – the relatively sudden dissemination of one-of-a-kind "accidental" cultural innovations like fire, sophisticated multi-component tools, personal adornments, rituals, art, shelter etc. Among the great apes, orangutans alone are reputed to display imitation of sophisticated skills ... often watching the keeper and picking locks or even paddling across a river in a canoe. If our species becomes extinct, they may well inherit the earth.

This type of gene-culture co-dependence suggests that the nature/nurture debate is meaningless in the context of human mental functions; it's like asking whether the wetness of water derives mainly from the H_2 or the O_2 that constitute H_2O. Our brains are inextricably bound to the cultural mileu they are immersed in and, if raised in a cave by wolves or in a culture-free environment (like Texas), we would barely be human—just as a single cell cannot exist without its symbiotic mitochondria. A Martian taxonomist watching the evolution of hominids would be struck by the observation that the behavioral difference (caused by culture) between post-twentieth-century *Homo sapiens* and early *Homo sapiens* (say 75,000 years ago—before the great leap forward) is actually much greater than the difference between *Homo erectus* and *Homo sapiens*! If he used behavioral criteria alone—rather than anatomy—he might classify the former two (late and early *sapiens*) as two different species and the latter two as a single one![10]

In chapter 2 I mentioned the "blindsight" syndrome, in which a patient with visual cortex damage cannot consciously see a spot of light shown to him but is able to use an alternative spared brain pathway to guide his hand unerringly to reach out and touch the spot. I would argue that this patient has a representation of the light spot in his spared pathway, but without his visual cortex he has no representation of the representation—and hence no qualia "to speak of." Conversely, in a bizarre syndrome called Anton's syndrome, a patient is blind owing to cortical damage but *denies* that he is blind. What he has, perhaps, is a spurious metarepresentation but no primary representation.[11] Such curious uncoupling or dissociations between sensation and conscious awareness of sensations are only possible because representations and metarepresentations occupy different brain loci and can therefore be damaged (or survive) independently of each other, at least in humans. (A monkey can develop a phantom limb but never Anton's syndrome or hysterical paralysis.) Even hypnotic induction in normal people can generate such dissociations—so-called "hidden observer" phenomena—leading to intriguing questions such as 'Can you hypnotically eliminate denial in an Anton's patient or demonstrate a form of blindsight after hypnotically inducing blindness in someone?'[12]

The flip side of this is, just as we have metarepresentations of sensory representations and percepts, we also have metarepresentations of motor skills and commands such as "waving goodbye," "hammering a nail in the wall" or "combing," which are mainly mediated by the supramarginal gyrus of the left hemisphere (near the left temple). Damage to this structure causes a disorder called ideomotor apraxia. Sufferers are not paralyzed, but, if asked to "pretend" to hammer a nail into a table, they will make a fist and flail at the tabletop (i.e., they do not mime the action accurately by holding an imaginary hammer shaft as a normal person would). Or when asked to mime combing her hair a patient will make a fist and bang it on her head, even though she understands the instruction and is perfectly intelligent in other respects. The left supramarginal gyrus is required for conjuring up an internal image—an explicit metarepresentation—of the intention and the complex motor–visual–proprioceptive "loop" required to carry it out. That the representation of the movement itself is not in the supramarginal gyrus is shown by the fact that if you actually give the patient a hammer and nail he will often execute the task effortlessly, presumably because with the real hammer and nail as "props" he doesn't need to conjure up the whole metarepresentation. (I have noticed that some of these patients even have difficulty with pointing, or looking at what's being pointed to, as though their sense of intentionality, of "aboutness" or "thatness," is to some extent compromised.)

For an act to be fully intentional the person has to be aware of—i.e., to anticipate—the full consequences of the act and must desire the consequences, as has been pointed out by the Oxford philosopher Anthony Kenny. (For example, if someone forces you to sign a document at gunpoint, you anticipate the signing of it but don't want to do it.) I suggest that anticipation and awareness are partly in the supramarginal gyrus and desire requires the additional involvement of the anterior cingulate and other limbic "emotional" structures. The sense of free will associated with the activity of these structures may be the proverbial carrot at the end of the stick that keeps goading the donkey in you into action.

Both a chimp and a human can reach out and grab a chocolate bar, but only a human can reflect on the long-term consequences and withhold the action because he or she is on a diet. (Intriguingly, patients with frontal lobe damage cannot do this withholding; they are not capable of "free won't," you might say. I would be very surprised if a frontal patient could go on a diet.) Patients with ideational and ideomotor apraxia have great difficulty in making judgments about whether other people's actions are intentional or not; they would probably make terrible judges or criminal lawyers. And a time may come when we may be able to do brain scans to determine whether a defendant is guilty of premeditated murder or merely of manslaughter (leading to new fields such as "neurojurisprudence" and "neurocriminology").

A crucial yet elusive aspect of self is its self-referential quality, the fact that it is aware of itself. As we noted earlier, once representations that initially served more "primitive" goals had evolved to a certain minimum level of sophistication, they became associated in the hominid brain, with metarepresentations intimately linked to one's sense of agency, personhood, and purpose (and the ability to juggle perceptual tokens "off line"—an important aspect of abstract thinking). The resulting "awareness that you are aware" or knowing that you know or "wanting to want" (for indeed not wanting to want) is what gives the "I" it's unique self-referential character.

Free will—the capacity to plan open-ended scenarios and try out even improbable scenarios entirely in the mind by juggling images and symbols—if linked with episodic memories, enables you to see yourself as an active agent doing things in the future (or past) and thereby generating a full-fledged sense of self. As a bonus, this ability would also enable you to present yourself to others as a predictable human being with certain stable attributes, an important capacity for intensely social creatures like us. By combining behavioral studies on patients with brain lesions—the main theme of my work—with functional brain imaging, and viewing the results from an evolutionary perspective, we can begin to elucidate these different components of self and finally tackle the mystery of how the components interact to generate awareness and self-consciousness.

The widely used phrase "raw awareness of sensations," or "primary awareness," employed by my colleagues to designate an earlier phylogenetic stage, is an oxymoron. "Awareness" simply doesn't mean anything without a metarepresentation—an awareness of awareness and a concomitant sense of self. If you are not aware that you are aware, then by definition, you are not aware! Humans are unique in this respect. The Victorian biologists Alfred Russel Wallace and Richard Owen were correct in asserting that a huge gap separates us mentally from other beasts. But, as I have suggested in this chapter contrary to Owen, what sets us apart from other mammals, including other primates, is not any single structure—such as the hippocampus minor—but a set of circuits that includes the temporo-parieto-occipital junction (especially the angular and supramarginal gyri), the Wernickes area (concerned with semantics) and the anterior cingulate with its limbic connections ("attention," "will," "desire," and the right parietal and insula concerned with embodienent). These structures are for consciousness what chromosomes and DNA were for heredity. Know how they perform their individual operations, how they interact, and you will know what it means to be a conscious human being.

And there I must close. As I said in chapter 1, my goal was not a complete survey of our knowledge of the brain, but I hope I've succeeded in conveying the sense of excitement that my colleagues and I experience each time we try to tackle one of these problems, be it synesthesia, hysteria, phantom limbs, free will, blindsight, neglect or any of the other syndromes I have mentioned. By studying these strange cases and asking the right questions, neuroscientists can begin to answer some of those lofty questions that have preoccupied philosophers since the dawn of history: What is free will? What is body image? Why do we blush? What is art? What is the self? Who am I? – questions that until recently were soley the province of metaphysics.

No enterprise is more vital for the well-being and survival of the human race. This is just as true now as it was in the past Remember that politics, colonialism, imperialism and war also originate in the human brain.

Notes

Chapter I: A Pain in the Brain

1 Hirstein and Ramachandran (1997); Ellis, Young, Quale and De Pauw (1997).

2 Ramachandran and Hirstein, 1998; Ramachandran, Rogers-Ramachandran and Stewart, 1992; Melzack, 1992. These experiments on phantom limbs were inspired, in part, by the pioneering physiological studies of Mike Merzenich, Patrick Wall, John Kaas, Tim Pons, Ed Taub and Mike Calford. Additional evidence for the remapping hypothesis comes from people who have undergone other types of sensory deprivation.

We have encountered two patients in whom after amputation of a leg, sensations were referred from the genitals to the phantom foot. One gentleman told us that even erotic sensations were referred from penis to leg so his orgasm was "much bigger than it used to be" (Ramachandran and Blakeslee, 1998). Perhaps some minor cross-wiring occurring even in normals might explain why feet are often considered erogenous zones and why we have foot fetishes. We prefer this anatomical view to Freud's far-fetched theory that foot fetishes occur because feet resemble the penis.

We predicted that after the fifth nerve (innervating the face) is cut the converse should be observed: sensations should be referred in a topographically organized manner from the face to the phantom. This has now been demonstrated by Stephanie Clarke (Clarke et al., 1996).

How massive an amputation is required to get the remapping to occur? Giovanni Berlucchi and Salvatore Aglioitti have shown that after amputation of an index finger a map of just that finger alone may be seen draped neatly across the face. And we have also seen referral from adjacent fingers. The referral was modality specific and topographically organized (Ramachandran and Hirstein, 1997).

In our first patient we had seen face-to-hand referral in four weeks and we suggested that the phenomenon was at least partly due to unmasking or activating previously silent connections that already existed between the face area and hand area rather than sprouting new axon terminals. In collaboration with David Borsook and Hans Beiter we found some degree of referral occurring less than 24 hours after deafferentation of an arm, implying that the unmasking idea must be at least partially correct. More recently there have even been intriguing reports of pressure cuff–induced anesthesia in an arm leading to sensations referred from face to hand, but this needs replication.

These referred sensations in amputees do not tell us whether the reorganization is occurring in the cortex or in the thalamus. Some years ago we suggested that the issue might be resolved by systematically looking for referred sensations in stroke patients who have zones of anaesthesia caused by lesions in touch pathways leading from thalamus to cortex. If referred sensations do occur in them (e.g., from face to arm) we can conclude that the reorganization is at least partly cortical. The existence of such referral has now been observed by several groups (Turton and Butler, 2001).

3 These observations on referred sensations also have the important implication that the sensory qualia of *location* (e.g., this sensation is coming from my hand, not my face) depends entirely on which part of the sensory cortex is activated—not on the actual source of the sensory stimulus—the face.

Yet I have found that if a person is born without arms this referral of touch from the face to the phantom does not occur, even though the subject does have a phantom; the stimulation of regions originally destined to report "hand" to higher centers must in these cases be reporting "face"! Similarly, Mriganka Sur has shown that if the visual pathways in a newborn ferret are redirected to the auditory cortex, viable connections are formed and the ferret now sees with its auditory cortex. How this reassignment of qualia labels occurs in people with congenitally missing limbs is a fascinating question (Hurley and Noe, 2003).

We are now trying to answer these questions by exploring the emergence (and precise topographic localization) of phosphenes (visual qualia) in patients with congenital vs. acquired blindness while we stimulate their visual areas artificially using transcranial magnetic stimulation.

4 These phantom movements originate because every time the motor center in the front of the brain sends a signal to the missing arm it sends a "copy" of the signal to the cerebellum and parietal lobes and these commands themselves are experienced

as movements even if there is no actual moving limb. Liz Franz, Rich Ivry and I showed this experimentally. Normal people cannot simultaneously do very dissimilar actions with their two hands—e.g., draw a circle with one hand and a triangle with the other. We found this was equally true if the patient "moved" his or her phantom to mime drawing a triangle—it interfered with the real hand's drawing, implying that commands to the phantom must be centrally monitored even if the arm is missing (Franz and Ramachandran, 1998).

5 These results obviously imply that the neural substrate of the body's image—in the parietal lobes—can be profoundly modified by experience. But there must also be an innately specified genetic template. We, and others, have found that some people born without arms experience vivid, complete phantom arms that even gesticulate and point to things.

It would be interesting to investigate male-to-female transsexuals from this point of view. A majority of patients undergoing amputation of the penis for cancer report feeling a vivid phantom penis and phantom erections. On the other hand, transsexuals report that "this appendage—this penis—doesn't feel like part of me. I have always felt like a woman trapped in a man's body," suggesting that this person's genetically specified brain-sex and corresponding body image is female rather than male. If so, one would predict a much smaller incidence of phantom penises after amputation of the organ in transsexuals than in "normal" adult men. Conversley, female to male transexuals may experience a phantom penis long before gendor reassigned surgeory, since their brain sex and body image is already male.

Intriguingly, some men with intact penises also report mainly having phantom erections rather than real ones (S. M. Anstis, personal communication).

6 Perhaps even the paralysis seen after a stroke is, in part, a form of learned paralysis that can be "cured" by a mirror. Preliminary results from our group (Altschuler et al., 1999) and by others (Sathian, Greenspan and Wolf, 2000; Stevens and Stoykov, 2003) have been encouraging, but systematic double-blind placebo-controlled studies are needed. The results would have tremendous implications even if only a small proportion of patients were helped by the procedure, given that 5 percent of the world's population will eventually suffer from stroke-related paralysis of an arm or leg.

The extent to which these techniques are practical in a clinical setting needs systematic evaluation, but meanwhile the general principles we established in the early nineties are here with us to stay. They can be stated as three propositions: First, there is a tremendous degree of latent plasticity in the adult brain; so much so that one should think of the sensorium as being in a state of dynamic equilibrium with—and being constantly updated by—the environment it is immersed in. Second, the idea that the brain consists of modules organized in a serial, hierarchical manner—with the output of each module being essentially complete before being delivered to the next level ("bucket brigade" model) needs to be discarded; there is a tremendous back and forth interplay of forces not only "vertically" across stages in the hierarchy, but also "laterally" across brain modules. Third—as a corollary of the first two—sensory inputs from one source eg vestibular (as we shall see) or visual, can profoundly influence—sometimes even substitute for a dysfunctional sensory system. The potential clinical and theoretical implications of these findings are obvious; they may represent a turning point in rehabilitation neurology. The work of Paul Bach Y Rita and—more recently—Alvaro PasquaLeone strongly vindicates this new view of brain function.

7 Our overall strategy has been the intensive investigation of neurological syndromes that have been mainly regarded as "oddities" in the past … whether phantom limbs, synesthesia or the Capgras delusion.

One problem is that both in neurology and in psychiatry many bogus syndromes have been described that represent little more than an attempt by a clinician to have a syndrome named eponymously. It can be difficult to decide which ones are likely to be genuine and worthy of study. For example there is a syndrome called De Clerambault's syndrome which is defined as "A young woman developing a delusional fixation that a much older, successful and famous man is passionately in love with her but doesn't realise it." Ironically there is no name for the syndrome in which an old man develops the delusion that a young woman is attracted to him but is in denial about it. Surely this latter syndrome is far more common (S. M. Anstis, personal communication), yet it remains unnamed! (A feminist might argue that this is because the vast majority of psychiatrists who make up names for syndromes are male.)

On the other hand, certain syndromes such as cortical color blindness (achromotopsia), motion blindness, the commissurotomy ("split brain") patients studied by Mike Gazzaniga, Joe Bogen and Roger Sperry and anterograde amnesia (studied by Brenda Milner, Elizabeth Warrington, Alan Baddeley, Larry Squire and others) have enormously enriched our understanding of the brain even though they were originally described as single case studies. Even the cellular and biochemical mechanisms underlying the physical changes that embody the "memory trace" have now been explored in intricate detail—culminating in a Nobel prize awarded to Eric Kandel.

Chapter 2: Believing Is Seeing

1 Bear in mind, though, that this is only an analogy. One key difference is that while driving I can voluntarily switch attention to my driving and ignore the conversation, but in the case of blindsight the information processed by the "unconscious" pathway cannot be accessed even through attention.

For detailed descriptions of blindsight see Weiskrantz, 1986 and Stoerig and Cowey, 1989. Some researchers—especially Semir Zeki—regard blindsight as not being a real phenomenon.

2 Ramachandran, Altschuler and Hillyer, 1997.

3 Ramachandran, 1995; Ramachandran and Blakeslee, 1998. In 1996, I proposed in the journal *Medical Hypotheses* that the extreme mood swings of bipolar disorder and "manic-depressive illness" may result from an actual alternation between the manic, delusional left hemisphere and the "depressed" right hemisphere. J. D Pettigrew (2001, personal communication)

has tested this notion and found that transcranial magnetic stimulation or caloric irrigation can shift the balance of relative activation between the two hemispheres and restore equilibrium of mood in the patient."

4 Frith and Dolan, 1997.

5 Ramachandran and Rogers-Ramachandran (1996).

6 In 1997 Eric Altschuler, Jamie Pineda and I showed that the blocking of mu waves in human EEG may provide an index of mirror neuron activity. Such blocking occurs when a normal individual voluntarily moves his hand *or* merely watches someone else's hand moving. Intriguingly, we found that in autistic children the blocking occurs for voluntary movements as in normals but not while watching someone else's hand—suggesting that they have a deficiency in their mirror neuron system that might partially explain their lack of empathy and impoverished "theory of other minds" (Altschuler, et al., 2000).

Or consider the first time a newborn infant reciprocates his mother's smile. This may require creating a complex internal representation of the visual characteristics of the mother's smile in the infant's brain and then "translating" this to be remapped onto motor maps representing the appropriate sequence of muscle twitches. This ability may depend crucially on mirror neuron or, more parsimoniously, it may be a stereotyped reflex response to the mother's smile. The only way to find out would be to determine whether (and at what age) a baby can first begin to mimic a NONstereotyped expression such as an asymmetric smile, a weird expression, or a wink. (And can monkeys and apes do this?) This would be a landmark—a turning point in both ontogeny and phylogeny of brain function.

It's also not clear whether the mere visual appearance of someone's grasping hand is enough to set off these neurons and cause mu wave suppression or whether you need to impute intention to the hand. What if you watched a grasping dummy hand that was obviously powered mechanically or an anesthetized, paralyzed hand that was passively opened and closed by mechanical pulleys? Would mu wave suppression occur?

7 As medical scientists we have two agendas: First, practical applications for treating patients, and second, theoretical insights into normal brain function. Our demonstration of extreme malleability of body image and modulation of phantom pain using mirrors, and our suggestion that the right hemisphere is sensitive to discrepancies or "anomalies," inspired the Australian biologist John Harris to propose an ingenious hypothesis that, just like nausea, pain *itself* is best conceived of as the organism's response to discrepancies in sensory input that activate the right hemisphere. If so, one wonders whether any type of chronic pain may benefit from asymmetric hemispheric activation through vestibular/caloric stimulation. This is especially likely for pain involving distortions of body image (such as complex regional pain syndrome type 1) but could conceivably also help relieve *any* kind of chronic pain that results from a disproportionate cerebral reaction to relatively minor injury. Or, who knows, perhaps it might alleviate even emotional "pain." Eric Altschuler has seen some preliminary hints of this.

Chapter 3: The Artful Brain

1 My book *The Artful Brain* is due to be published in 2005. Also see the website of Bruce Gooch (University of Utah) on the laws of art: http://www.cs.utah.edu/~bgooch/.

2 Experiments dating back to Francis Galton show that averaging several faces together often produces a face that is quite attractive. Does this contradict my peak shift law? Not necessarily. Averaging probably works by eliminating minor blemishes and distortions such as warts, disproportionate face parts, asymmetries, etc., which makes evolutionary sense.

Yet the peak shift principle would predict that the most attractive female face is not necessarily the "average" but usually one with the right kind of exaggeration. For example, if you subtract the average female face from the male and amplify the difference you would end up with an even more gorgeous face—a "superfemale" with neotonous features (or a male stud-muffin with exaggerated jawline and eyebrows).

3 Just for fun, let's see how far we can take this argument. Cubism involves taking the usually invisible other side of an object or face and moving it forward to the same plane as the side that is visible: two eyes and two ears visible on the profile view of a face, for example. This has the effect of liberating the observer from the tyranny of a single viewpoint: you don't have to walk around the object to see its other side. Every art student knows this is the gist of Cubism but few have raised the question of why it is appealing. Is it just shock value, or is there something else?

Let us consider the response of single neurons in the monkey brain. In the fusiform gyrus individual neurons often respond optimally to a particular face, e.g., one cell might respond to the monkey's mother, one to the big alpha male and one to a particular side-kick monkey—a "Phanka waala cell," you might say. Of course the one cell doesn't "contain" all the properties of the face; it is part of a network that responds selectively to that face, but its activity is a reasonably good way of monitoring the activation of the network as a whole. All this was shown by Charlie Gross, Ed Rolls and Dave Perrett.

Interestingly, a given neuron (say an "alpha male face neuron") will respond only to *one* view of that particular face—e.g., its profile. Another one nearby might respond to semi-profile and a third one to a full frontal of that face. Clearly, none of these neurons can by itself be signaling the concept "alpha male" because it can respond only to one view of him. If the alpha male turns slightly the neuron will stop firing.

But at the next stage in the visual processing hierarchy you encounter a new class of neurons that I'll call "master face cells" or "Picasso neurons." A given neuron will respond only to a particular face, e.g., "alpha male" or "mother," but unlike the neurons in the fusiform the neuron will fire in response to *any* view of that particular face (but not to any other face). And that, of course, is what you need for signaling: "Hey—it's the alpha male: watch out."

How do you construct a master face cell? We don't know, but one possibility is to take the outgoing wires—axons—of all the "single viewpoint" cells in the fusiform that correspond to a single face (e.g., alpha male) and make the axons converge on to a single master face cell—in this case the alpha male cell. As a result of this pooling of information you can present any view of the alpha male and it will make at least one of the individual view cells in the fusiform fire, and that signal will in turn activate the master cell. So the master cell will respond to any view of that face.

But now what would happen if you were simultaneously to present two ordinarily incompatible views of the face in a single part of the visual field in a single plane? You would activate two individual face cells in parallel in the fusiform and hence the master cell downstream will get a double dose of activation. If the cell simply adds these two inputs (at least until the cell's response is saturated), the master cell will generate a huge jolt, as if it were seeing a "super face." The net result is a heightened aesthetic appeal to a Cubist representation of a face—to a Picasso!

Now the advantage of this idea—however far-fetched—is that it can be tested directly by recording from face cells at different stages in the monkey brain and confronting them with Picasso-like faces. I may be proved wrong, but that is its strength—it can at least be *proved* wrong. As Darwin said, when you close one path to ignorance, you often simultaneously open a new one toward the truth. This cannot be said for most philosophical theories of aesthetics.

4 If these arguments about "aesthetic universals" are correct then an obvious question arises: why doesn't everyone like a Picasso? The surprising answer to this question is that everyone *does*, but most people are in denial about it. Learning to appreciate Picasso may consist largely in overcoming denial!

(Just as the Victorians initially denied the beauty of Chola bronzes until they overcame their prudishness.) Now I know this sounds a bit frivolous, so let me explain. We have known for some time now that the mind isn't one "thing"—it involves the parallel activity of many quasi-independent modules. Even our visual response to an object isn't a simple one-step process—it involves multiple stages or levels of processing. And this is especially true when we talk about something as complex as aesthetic response ... it is sure to involve many stages of processing and many layers of belief. In the case of Picasso I would argue that the "gut level" reaction—the "a-ha" jolt—may indeed exist in everyone's brain, caused, perhaps, by early limbic activation. But then in most of us higher brain centers kick in telling us, in effect, "Oops! That thing looks so distorted and anatomically incorrect that I had better not admit to liking it." Likewise, a combination of prudery and ignorance might have vetoed the Victorian art critics' reaction to voluptuous bronzes—even though neurons at an earlier stage are firing away, signaling peak shifts. Only when these subsequent layers of denial are peeled off can we begin to enjoy a Picasso or a Chola. Ironically, Picasso himself derived much of his inspiration from "primitive" African art.

5 In my book *Phantoms in the Brain* I suggested that many of these laws of aesthetics—especially peak shift—may have powerfully influenced the actual course of evolution in animals, an idea that I call the "perceptual theory of evolution." A species needs to be able to identify its own species in order to mate and reproduce, and to do so it uses certain conspicuous

perceptual "signatures"—not unlike the gull chick pecking the stick with three stripes. But because of the peak shift effect (and ultranormal stimuli) a mate might be preferred that doesn't "resemble" the original. In this view the giraffe's neck grew longer not merely to reach tall acacia trees but because giraffes' brains are wired to automatically show greater propensity to mate with more "giraffe-like" mates, i.e., mates with the giraffe trait of longer necks. This would lead to a progressive caricaturization of descendants in phylogeny. It also predicts less variation in the externally visible morphology and colors in creatures which don't have well-developed sensory systems. (e.g., cave dwellers) and less florid variations of internal organs which cannot be seen.

This notion is similar to Darwin's idea of sexual selection—i.e., peahens preferring peacocks with larger and larger tails. But it is different in three respects.

My argument, unlike Darwin's, doesn't apply only to secondary *sexual* characteristics. It argues that many morphological features and labels identifying species (rather than sexual) differences might propel evolutionary trends in certain directions.

Although Darwin invokes "liking larger tails" as a principle in sexual selection he doesn't explain *why* this happens. I suggest that it results from the deployment of an even more basic psychological law wired into our brains that initially evolved for other reasons, such as facilitating discrimination learning.

Given our seagull chick principle, i.e., the notion that the optimally attractive stimulus need not bear any obvious surface resemblance to the original (because of idiosyncratic aspects of neural codes for perception), it is possible that new trends in morphology will start that have no immediate functional significance and may seem quite bizarre. This is different from the currently popular view that sexual selection of absurdly large tails occurs because they are a "marker" for the absence of parasites. For example, certain fish are attracted to a bright blue spot applied by the experimenter on a potential mate, even though there is nothing remotely resembling it in the fish. I predict the future emergence of a race of blue spotted fish even though the blue spot is not a marker of sex or of species, or an advertisement for good genes that promote survival. Or perhaps a race of seagulls with striped beaks!

Note that this principle sets up a positive feedback between the observer and the observed. Once a "species label" is wired into the brain's visual circuits, then offspring who accidentally have more salient labels will survive and reproduce more, causing an amplification of the trait. That in turn will make the trait an even more reliable species label, thereby enhancing the survival of those whose brains are wired up to detect the label more efficiently. This sets up a progressive gain amplification.

6 Another way to test these ideas would be to obtain a skin conductance response (SCR) which measures your gut-level emotional reaction to something by measuring increases in skin conductance caused by sweating. We know that familiar faces usually evoke a bigger response than unfamiliar ones—because of the emotional jolt of recognition. The counterintuitive prediction would be that an even bigger response would be shown to a caricature or Rembrandt-like rendering of a familiar face than to a realistic photo of the same face. (One could control for the effects of novelty caused by the exaggeration by comparing this response to that elicited by a randomly distorted familiar face or an "anticaricature" that reduced rather than amplified the difference.)

I am not suggesting that an SCR is a complete measure of a person's aesthetic response to art. What it really measures is "arousal," and arousal doesn't always correlate with beauty—it only implies "disturbing." Yet few would deny that "disturbing" is also part of the aesthetic response: just think of a Dalí or Damien Hirst's pickled cows. This is no more surprising than the fact that we seem to, paradoxically, "enjoy" horror movies or white-knuckle rides. Such activity may represent a playful rehearsal of brain circuits for future genuine threats and the same could be said of visual aesthetic responses to disturbing, attention-grabbing, visual images. It's as if anything salient and attention-grabbing—almost by its very nature—encourages you to look at it more to process it further, thereby fulfilling at least the first requirement of art. But the "attention-grabbing" component would be the same for the randomly distorted face and the caricature, whereas only the latter will have an additional component added by the peak shift. These different "components" of the aesthetic response will eventually be dissected more as we develop a clearer understanding of the connections between the visual areas and limbic structures and of the logic that drives them (the "laws" we have been discussing). So a randomly distorted nude might excite only the amygdala ("interest + fear") whereas the peak-shifted Chola bronze would excite the amygdala (interest) *and* the septum and nuclus accumbens (adding "pleasure" to the cocktail, so you end up with "interest + pleasure").

An analogy with IQ tests might be illuminating. Most people would agree that it is ludicrous to measure something as multidimensional and complex as human intelligence using a single scale such as IQ. Yet it's better than nothing if you are in a hurry: trying to recruit sailors, for example. An individual with an IQ of 70 is unlikely to be bright by any standard and one of IQ 130 is unlikely to be stupid.

In a similar vein I would suggest that even though the SCRs can provide only a very crude measure of aesthetic response, better a crude measure than none. And it can be especially useful if combined with other measures such as brain imaging and single neuron responses. For example, a caricature or a Rembrandt might activate face cells in the fusiform more effectively than a realistic photo.

It may be helpful, also, to make a further distinction between "aesthetic universals" vs. "art"—which is in some ways a more loaded term. Aesthetic universals would include so-called design, but wouldn't include pickled cows.

7 It isn't clear what "kitsch" is, but unless we tackle this we cannot really claim to have completely understood art. After all, kitsch art also sometimes deploys the same "laws" I am talking about—e.g., grouping or peak shifts. So one way of finding out what neural connections are involved in "mature aesthetic appreciation" would be to do brain imaging experiments in which you subtract the subject's reaction to kitsch from her reaction to high art.

One possibility is that the difference is entirely arbitrary and idiosyncratic, so one man's high art might be another's kitsch. This seems unlikely, since we all know that you can evolve from appreciating kitsch to appreciating the genuine thing, but you can't slide backwards. I would suggest instead that kitsch involves merely *going through the motions* of applying the laws we have talked about, without a genuine understanding of them. The result is "pseudo art" of the kind found in hotel lobbies in North America.

As an analogy we can compare kitsch to junk food. A strong solution of sugar elicits a gustatory jolt, as every child knows, and powerfully activates certain taste neurons. This makes sense from an evolutionary standpoint: our ancestors (as Steve Pinker points out) often had to go on carbohydrate binges in preparation for enduring frequent famines. But such junk food cannot begin to compete with gourmet food in producing a complex multidimensional titillation of the palate (partly because of reasons divorced from the original evolutionary functions, e.g., peak shift and contrast, etc. applied to taste responses and partly to provide a more balanced meal that's more nutritious in the long run). Kitsch, in this view, is visual junk food.

8 Do animals have art? Some of these universal laws of aesthetics (e.g., symmetry, grouping, peak shift) not only may hold across different human cultures but may even cross the species barrier. The male bower bird is quite a drab fellow but an accomplished architect and artist, often building enormous colorful bowers—the avian equivalent of a bachelor pad; a sort of Freudian compensation for his personal appearance, you might say. He makes elaborate entryways, groups berries and pebbles according to color similarity and contrast, and even collects shiny bits of cigarette-foil "jewelery." Any of these bowers could probably fetch a handsome price if displayed in a Fifth Avenue gallery in Manhattan and falsely advertised as a work of contemporary art.

The existence of aesthetic universals is also suggested by the fact that we humans find flowers beautiful, even though they evolved to be beautiful to bees and butterflies, which diverged from our ancestors in Cambrian times. Also, principles such as symmetry, grouping, contrast and peak shift used by birds (e.g., birds of paradise) evolved to attract other birds, but we are similarly moved by them.

In response to this chapter, Richard Gregory and Aaron Schloman have pointed out to me that if such laws exist it might be possible to program at least some of them into a computer and thereby generate visually appealing pictures. Something along these lines was in fact attempted by Harold Cohen many years ago at UCSD and his algorithms do indeed produce attractive pictures that fetch handsome prices.

9 Not all Western art critics were as obtuse as Sir George. Listen to the French scholar Renée Grousset describing the Shiva Nataraja (Figure 3.4):

"Whether he be surrounded or not by the flaming aureole of the Tiruvasi—the circle of the world, which he both fills and oversteps—the king of the dance is all rhythm and exaltation. The tambourine, which he holds with one of his right hands, calls all creatures into this rhythmic motion, and they dance in his company. The conventionalized locks of flying hair and the blown scarf tell of the speed of this universal movement, which crystallizes matter and reduces it to powder in turn. One of his left hands holds the fire which animates and devours the world in this cosmic whirl. One of the god's feet is crushing a titan, for 'this dance is upon the bodies of the dead,' yet one of the right hands is making the gesture of reassurance (*abhayamudra*), so true it is that, seen from the cosmic point of view..., the very cruelty of this universal determinism is kindly, and the generative principle of the future. And, in more than one of our bronzes, the king of the dance wears a broad smile. He smiles at death and life, at pain and at joy alike, or rather, if we may be allowed so to express it, his smile *is* both death and life, both joy and pain ... From this lofty point of view, in fact, all things fall into their place, finding their explanation and logical compulsion ... The very multiplicity of arms, puzzling as it may seem at first sight, is subject in turn to an inward law, each pair remaining a model of elegance in itself, so that the whole being of the Nataraja thrills with a magnificent harmony in his terrible joy. And as though to stress the point that the dance of the divine actor is indeed a force (*lila*)—the force of life and death, the force of creation and destruction, at once infinite and purposeless—the first of the left hands hangs limply from the arm in a careless gesture of the *gajahasta* (hand as the elephant trunk). And lastly as we look at the back view of the statue, are not the steadiness of these shoulders that support the world, and the majesty of the Jove-like torso, as it were a symbol of the stability and immutability of substance, while the gyration of the legs in their dizzy speed would seem to symbolize the vortex of phenomena."

Chapter 4: Purple Numbers and Sharp Cheese

1 In many "lower synesthetes" not only numbers but also letters of the alphabet—what we call graphemes—evoke specific colors. It is very likely that the visual appearance of a letter is also represented in the fusiform, so the "cross-activation" hypothesis can explain this as well.

In others it is the sound of the letter—the phoneme—that determines the color, and this may be based on cross-activation at a higher stage, near the TPO junction and angular gyrus (Ramachandran and Hubbard, 2001a, b).

2 Another finding that we made in 2002 (Ramachandran and Hubbard; Psychonomic society abstracts 7:79) also lends strong credibility to the view that in "lower synesthetes" (e.g., our subject JC) the colors evoked are genuine sensations that result from activity in the sensory regions of the brain. We presented a computer display of 5s in which there was a hidden triangle composed of 2s, as in Figure 1.8. This whole display was flashed briefly (for a third of a second) in the first frame of a movie and then replaced by a display in which the embedded triangle had shifted horizontally but the elements defining it and the background were randomly jittered. Normal volunteers just see random incoherent motion as the two frames are alternated; each element jumping to its nearest neighbor, but synesthetes like JC saw a triangle jumping horizontally left and right (guessing direction correctly on almost 100 percent of random sequences of trials). Normals performed little better than chance. This shows that colors that are evoked purely internally in the brain and don't exist in the display can nevertheless drive motion detecting mechanisms in the brain. Similar observations have been made more recently by M Palmieri and colleagues at Vanderbilt University.

3 This theory should not be taken to imply that learning in early childhood plays no role in synesthesia. In a sense it *must* play a role, given that we are not born with number-signaling neurons in the brain. So the cross-activation merely provides a substrate—it confers a propensity to link numbers with colors—but doesn't determine which number evokes which color.

Hence it is not surprising that the same number can evoke different colors in different synesthetes. Yet the distribution isn't completely random across synesthetes—"0" tends to be white far more than green, for example. Similarly, the correspondence between specific phonemes and colors may initially seem arbitrary, but if we classify phonemes into bilabial, dental alveolar, palatal, velar, labio dental (and voiceless vs. voiced, nasal, etc.) depending on their mode of representation in motor maps, certain correlations and patterns may emerge. Let us not forget the lesson from the periodic table of elements. The elements seemed to form clusters (e.g., halogens or alkaline metals) but no clear pattern could be discerned until Mendeleev discovered the "atomic number rule" that later resulted in the periodic table.

4 Additional evidence for this view comes from looking at the effects of changing the contrast of the numbers. In lower synesthetes, as the contrast is reduced the color becomes less and less saturated until below 8 percent contrast it vanishes completely, even though the number itself is still clearly visible (Ramachandran and Hubbard, 2002). This high level of sensitivity to physical stimulus parameters such as contrast points to cross-wiring at early stages in neural processing. What happens when the subject visualizes or imagines the number in front of him? Oddly enough we found that many subjects report the color to be more vivid. To understand this we have to bear in mind that when you imagine an object there is partial activation of the same sensory pathways in the brain as when that object is actually seen. This "top down" internally generated activation may be sufficient to cross-activate the color nodes. But when you actually look at a black number there is simultaneous activation of neurons in the brain that signal black, and they partially veto the synesthetic color. For a number representation evoked internally through imagery this vetoing doesn't occur, hence the color becomes more vivid.

5 Another relatively common type of synesthesia, described by Francis Galton, is the "number line." When asked to visualize numbers, such synesthetes see each number always in a particular location with different numbers (sometimes up to 30 or even 100) arranged sequentially along a line. Usually the line is highly convoluted—sometimes even doubling back on itself so that (say) 9 might be nearer to 2 than to 8 in Cartesian space.

We recently devised a technique for showing this objectively (Ramachandran and Hubbard, 2001b). When normal people are asked to say which of two numbers is larger, their reaction increases linearly with the numerical "distance" between the numbers—as if they were reading them off a perfectly straight number line—so that numbers closer together become harder to tell apart. (This was shown by Stanislas Dehaene.) But when we tested our synesthetes who had convoluted number lines that doubled back, we found this wasn't true. The reaction time no longer varied with the numerical distance alone—there seemed to be some sort of compromise between the Cartesian distance and numerical distance (Ramachandran and Hubbard, 2002).

There is another curious twist to this number form. We asked another subject who experienced a convoluted—but mainly horizontal—number line to lean rightward and tilt her head in the coronal plane to the right so that her ear touched her right shoulder. Remarkably, her number line also rotated 90 degrees along with her head, so it was now vertical with respect to gravity (and the individual numbers also rotated correspondingly). This observation implies that at least in this subject, the Galtonian number line is computed and represented in retinocentric coordinates—prior to vestibular correction for head tilt, rather than in an abstract world-centered coordinate system. We have previously shown that this is also true for other visual phenomena, such as shape from shading. Intriguingly, if she stood next to a wall the portion of the number line that went "through" the wall or obstacle became "fuzzy"—almost invisible. Putting a mirror instead of a wall next to her produced acute anxiety: "Those numbers are on my right but that means I ought to see them on my left . . .," reminiscent of mirror agnosia described in chapter 2. These experiments were done with my students Shai Azoulai and Ed Hubbard.

One wonders if the astonishing computational skills of autistic "number savants" are based on similar shortcuts taken through "wormholes" connecting different regions of their number landscapes. One of our subjects (Danielle) who had an extraordinarily convoluted number line seemed to make more errors with subtraction and division across points of sudden inflection than across other smoother regions. Other subjects have described to us how they could "wander the number landscape" allowing them to uncover hidden relationships between numbers "inspected" from novel vantage points. (Such claims are sometimes also made by true mathematical geniuses—not just savants). Other synesthetes tend to classify all letters as male or female or good or evil—a quirky manifestation of the universal human propensity to dichotomize the world to simplify processing. Such observations are intriguing but they also remind us how little we really know about the brain.

6 The *directionality* of synesthesia also requires comment. Many have noted that numbers evoke colors but colors rarely evoke numbers (though see below). Perhaps the manner in which "color space" is represented in maps in the brain vs. the manner of representation of graphemes confers an automatic bias toward unidirectional cross-activation (Ramachandran and Hubbard, 2001b).

7 Contrary to popular wisdom, *metaphors* in ordinary language are not arbitrary either—certain directions are preferred (Lakoff and Johnson, 1999) and this supports our claim about the analogy between metaphors and synesthesia. For example, we speak of a "loud shirt" but not of a "red sound"; a "soft" or

"rough" sound but not a "loud" texture. And we say "sharp taste" but never "sour touch." All of which, we suggest, may reflect anatomical constraints.

We have encountered only one synesthete who sees numbers when she sees colors but not the other way around. Indeed when she sees (say) a polka-dotted or check shirt composed of two colors, she instantly sees the *sum* of the two numbers and then decomposes it to realize that she has unconsciously added them. Such examples remind us that we are not dealing with physics here but with biology, where exceptions abound.

Synesthesia can also be used as a mnemonic aid. Many have told us how their color associations helped them learn to type (or learn musical scales) more rapidly than their peers because the letters (or notes) were "color coded" (Ramachandran and Hubbard, 2001a).

What about more exotic forms of synesthesia such as touch to taste (as in Cytowick's (2002) celebrated "Man who tasted shapes" or in our subject Matt Blakeslee). We have suggested this probably reflects the close proximity of the insular cortex which receives taste and the somatosensory hand area in the Penfield map.

Brain maps that are already partially connected are more likely to become involved in synesthetic cross-activation. Such maps are often close together anatomically (like the color and number areas in the fusiform, the color and hearing areas near the temporo-parieto-occipital (TPO) region or the touch/taste maps in the insula). But they don't have to be. Jamie Ward has recently studied synesthetes in whom phonemes evoke tastes and he implicates connections between the insula and Broca's area.

Do normal people also experience synesthesia? We all speak of certain smells—such as nail polish—being sweet, even though we have never tasted it; this might involve the close neural links and cross-activations between smell and taste—you can think of this as a form of synesthesia that exists in all our brains. This would make sense not only functionally—e.g., fruits are sweet and also smell "sweet," like acetone—but also structurally: the brain pathways for smell and taste are closely intermingled and project to the same parts of the frontal corex. A curious form of acquired (rather than congenital) synesthesia was studied recently by me and Shai Azoulai in a patient with blindness caused by long-standing damage to the visual pathways. When the patient moved his hand in front of his eyes in complete darkness he not only felt, but actually saw his hand. We suggested that this results from "top down" activation of visual centers by feedback from polymodal cells in the frontal and parietal. In a normal person this feedback would be vetoed by nulling signals from the visual pathways, but if the pathways are damaged, the nulling does not occur and the patient sees his hand.

Lastly, consider the fact that even as infants we scrunch up our noses and raise our hands when we encounter disgusting smells and tastes. Why is it that all cultures use the same word, "disgusting," and make the same facial expression for a person who is morally offensive? Why the same word as to describe a horrible taste? (Why not describe an offensive person as "painful," for example?) I suggest that once again this is because of evolutionary and anatomical constraints. In lower vertebrates certain regions of the frontal lobes have maps for smell and taste, but as mammals became more social the same maps become usurped for social functions such as territorial marking, aggression and sexuality, eventually culminating in mapping a whole new social dimension —morality. Hence the interchangeable words and facial expressions for olfactory/gustatory and moral disgust (Ramachandran and Hubbard, 2001a, b).

8 Are there neurological disorders that disturb metaphor and synesthesia? This has not been studied in detail but, as noted in the lecture, we have seen disturbances in the booba/kiki effect as well as with proverbs in patients with angular gyrus lesions. I tested one patient with anomia caused by a left angular gyrus lesion recently and found that he got fourteen out of fifteen proverbs wrong—usually interpreting them literally rather than metaphorically.

These findings challenge the commonly held view that "pragmatic" aspects of language are mediated exclusively by the right hemisphere (a view that was, in any event, derived from nonspecific laterality data that emphasizes hemispheric specialization with no information on actual lesion location). One possibility is that the left angular gyrus is involved in abstract "conceptual" metaphors and cross-modal metaphors like sharp cheese, whereas the right angular gyrus may be involved in "embodied" and spatial metaphors such as "The news weighed heavily on him," but this remains to be explored. There may be an additional role for the left supramarginal gyrus in action-based metaphor (e.g., "he reached for the stars" and "get a grip on yourself"). We recently examined a patient with a left supramarginal gyrus lesion and found that he not only had difficulty miming skilled actions as expected ("ideomotor apraxia"), but also had difficulty understanding action-based metaphors, usually interpreting them literally. We suggest that human homologues of the mirror-neuron system in this region may be involved in the representation of action-metaphors. My student Lidsay Oberman and I have obtained some preliminary evidence for this by measuring mu wave suppression when normal subjects listen to action metaphors.

Schizophrenics are also bad at interpreting proverbs but curiously they often unwittingly generate puns and "clang associations." I am stuck by the similarities between schizophrenics and patients with right fronto-parietal lesions. Delusions ("I am Napoleon," "I am not paralyzed") and hallucinations ("My left hand is touching your nose") occur in both and, like schizophrenics, patients with right-hemisphere lesions come up with puns and facetious humor. This suggests that they may have right fronto-parietal hypofunction and some abnormal excess of activity in the left.

Matha Farrah and Steve Kosslyn have shown that the generation and control of internally generated imagery is mainly a left-frontal function and I have suggested that "checking" it against reality is done in the right (Ramachandran and Blakeslee, 1998). So the lesions we are postulating here in schizophrenia would lead to uncontrolled, unchecked imagery (hallucinations) and beliefs (delusions).

9 A word is much more than a mere label. This was brought home to me vividly when I recently examined an Indian

patient with a language disorder called anomia, or "tip of the tongue" phenomenon, caused by damage to his left angular gyrus. In addition to his anomia (difficulty naming objects shown to him and in finding the right words during spontaneous speech) he had other symptoms of Gerstmann's syndrome: finger agnosia (inability to name fingers, neither his own nor the physician's) and left/right confusion (intriguingly he couldn't point to which shoe went with which foot even when actually seeing the shoes and feet simultaneously —a form of "chirality blindness").

When shown an object he would often produce semantically related words—e.g., when shown glasses he said "Eyesight medicine"—supporting the standard view that he *knew* what it was but the name label eluded him. However, there were many categories of objects for which this wasn't true. When shown a statue of the Indian god Krishna (whom any Indian child can identify instantly) he got it wrong, saying, "Oh, he's the god who helps Rama cross the ocean" (meaning the monkey god, Hanuman). When I then hinted "The name starts with Kr ..." he said, "Oh, of course, it's Krishna ... he *doesn't* help Rama." The same was true for a number of other objects that he initially miscategorized; correcting the name enabled him also to evoke correctly the appropriate semantic associations. The observation implies that, contrary to popular wisdom, a name is not merely a label; it's a magic key that opens up a treasury of meanings associated with what you are looking at.

Given that my patient couldn't name fingers, I wondered what he would do if I made the rude sign of "giving him the finger" (extended middle finger). He said I was pointing to the ceiling ... suggesting, again, that what is lost is not only a word name but even highly salient associations.

Lastly, he was also terrible with metaphors: this was the man who interpreted fourteen out of fifteen proverbs literally instead of metaphorically (see note 8), despite being perfectly normal in other intellectually demanding tasks. (This supports my idea that the left TPO—especially the angular gyrus—may have played a pivotal role in the emergence of metaphor in humans.)

10 If there is anything to this view then why don't all languages use the same word for the same object? After all we say "dog" in English, "chien" in French and "nai" in Tamil.

The answer is that our principle applies only to the ancestral proto-language when things were just getting off the ground. Once the basic framework was in place arbitrary differences emerged as languages diverged: a switch to the Sassurian mode. It's getting things started that's often a problem in evolution.

Support for this view also comes from studies in comparative linguistics (Berlin, 1994). A certain South American tribe has several dozen words for different fish species and an equally large number of bird names. If an English speaker is asked to classify these words—incomprehensible to him – into birds vs. fish, he scores well above chance. This points to a non-arbitrary link between object appearance and the sound used to denote it.

11 The syntactical "nesting" of clauses bears a striking resemblance to the execution of arm movements. If I say "touch your nose" you effortlessly move your hand along the shortest trajectory, adjusting the elbow angle, fingers, etc., using the appropriate sequence of muscle twitches. But you could also, if you wanted to, move your hand behind your neck, curving it forward to touch your nose, even though you have never done this before. So it's the goal (nose touching) and overall strategy (contract proximal muscles first and progressively "nest" other subroutines for more and more distal joints) alone that are specified—not the precise sequence of muscle twitches. This goal-directed "nesting" of motor subroutines is not unlike the embedding of clauses within larger sentences.

We also have to distinguish the question of the functional autonomy of syntax and semantics in the modern brain from the question of evolutionary origins. Syntax is almost certainly modular in a modern human because we know that patients with damage to the Wernicke's area can have one without the other. They can produce grammatically flawless but meaningless sentences (such as Chomsky's fictitious example "Colorless green ideas sleep furiously"), implying that the isolated Broca's area can generate syntactic structure on its own. But it does not follow from this that syntax didn't evolve from some preceding ability.

As an analogy, consider the three little bones in your middle ear used for amplifying sounds. It's a defining characteristic of mammals—our reptilian ancestors didn't have them. It turns out, though, that reptiles have three bones on each side of their multihinged lower jaw—suitable for swallowing large prey but not for chewing—whereas mammals have only one, the mandible. From studies in comparative anatomy we now know that because of their fortuitous anatomical location the two posterior bones in the jaws of reptiles became assimilated into the mammalian ear for hearing.

In modern mammals hearing and chewing are "modular" —independent of each other both structurally and functionally (i.e., you can lose your jawbone without becoming deaf). Yet once the evolutionary sequence is spelled out it becomes crystal clear that one function evolved from the other. And in my view the same sort of thing could have happened time and again in the emergence of syntax and other language capacities as outlined here, an idea that is abhorrent to many linguists.

One reason for tension between "pure" linguists and neuroscientists is that the former group is interested *only* in rules intrinsic to the system—not how and why the rules came to be, or how they are enshrined in neural architecture, and how they are related to other brain functions. To an orthodox linguist such questions are as meaningless as they would be for a number theorist interested in prime numbers, Fermat's theorem or Goldbach's conjecture. (And any talk of evolution or of neurons or the role of the angular gyrus in numerical skills would seem remote from his interests!) The key difference is that syntax evolved over 200,000 years or longer, through natural selection, whereas number theory is less than 2,000 years old and its intrinsic rules were neither selected for nor adaptive in any sense. Indeed, it's their very uselessness that makes them so seductive to many pure mathematicians!

Chapter 5: Neuroscience—The New Philosophy

1 Another possibility is that this "delay" has no function and occurs because of an inevitable smearing of neural events in space-time. Since there is no "cinema screen" in the brain being watched by a little person (homunculus) in real time, there's no reason to expect a precise synchrony between one's sense of volition and the neural cascades that generate the corresponding movements. This view has been eloquently championed by the eminent American philosopher Dan Dennett, and it has the advantage of parsimony. (Although the parsimony rule can be misleading in biology, given the way evolution works; as Crick once said: God is a hacker—not an engineer.) Wegner (••••) and Churchland (••••, ••••) have made important contributions to the problem of free will. Thanks to them, and to Francis Crick, Kristoff Koch and Gerald Edelman, the study of consciousness is now considered respectable.

One difficulty I have with the time-smearing notion is why the error in judging synchrony between brain events and the feeling of will is systematic and always in one direction; if it is really just "error" you might expect to feel the willing at random points of time clustered around the brain events.

In general, it's fair to say that philosophers have made very little progress in understanding consciousness, but there are a few exceptions, notably Pat and Paul Churchland, John Searle, Dan Dennett, Jerry Fodor, David Chalmers, Bill Hirstein, Ned Block, Rick Grush, Alva Noe and Susan Hurley (although even these enlightened few tend to take perverse delight in disagreeing among themselves ad nauseam—an occupational hazard).

2 If these ideas are right, then we can also make another prediction. A normal person sending a command to move an arm receives feedback from vision and proprioception (joint and muscle sense) that the arm is obeying the command. But using a system of mirrors and a hidden assistant wearing a glove it is possible to make someone see his arm as perfectly stationary—i.e., *not* moving. Even though the motor *commands* are being monitored by his brain and the arm is *felt* to be moving, it is *seen* as stationary (Ramachandran and Blakeslee, 1998). Normal people experience a powerful jolt when confronted with this incongruity, saying things like, "My God, what's going on! Why isn't the arm moving?" But when we tried this on the non-paralyzed right hand of a patient with anosognosia caused by right-hemisphere damage, she calmly reported that she could see the arm moving perfectly well—she ignored the mismatch. Pursuing the analogy between anosognosia and schizophrenia further, I predict that schizophrenics will do the same thing when confronted with this type of mirror box—they will hallucinate their arm moving.

3 There are actually two different versions of the qualia problem (Ramachandran and Blakeslee, 1998). The first one—which in my view is intimately linked to one's sense of self—is the puzzle of why there should be any subjective sensations at all. Why can't everyone, including me, be zombies going about their business? Why are there two parallel descriptions or "stories" about the world—the subjective "I" story and the objective "It" story? The second problem is why the sensations take the particular form that they do. This second is, in my view,

more tractable by the methods of science and its solution may take us closer to also answering the first question.

The first question can be illustrated by the following paradox. Imagine I show you two completely identical human beings—one (without his knowledge) condemned from this moment on to live in a cave and be tortured and the other outside, perpetually enjoying himself. If I ask *you*, "Would it be OK if I swapped the two people while they were asleep?" you would say fine, or at least you wouldn't see any particular reason why they shouldn't be. But if I now modify the question and say, "Assume one of them (the one outside the cave) is you ... would it be OK if I did the swap?" you would say, "No ... don't." Yet how can you logically justify this if you believe that only an "objective world"—a third person account—exists? A question similar to this was raised in Sankhya philosophy in ancient India (as quoted by Erwin Shrödinger in "Mind and matter").

As an example of the second question—about qualia—consider the manner in which we experience two different physical dimensions: wavelength (in vision) and pitch (in sound). Even though wavelength is a continuous dimension we experience colors as four qualitatively distinct sensations—red, yellow, green and blue. These four seem subjectively "pure"—they do not seem to be made of other colors or intermediate between other colors. Adjacent colors in this set of four are "miscible," i.e., we can see orange as being a blend of red and yellow and purple as comprising red and blue. But non-adjacent ones are immiscible, like oil and water—it is hard even to imagine a bluish yellow or a reddish green. Thus color sensations seem "chopped up" into four immiscible bits. But this isn't true of the frequency of sound waves: we hear the full range from very low to very high pitch as a continuum, with no breaks in qualia.

All this is obvious, but the question is why it should be so. To say that it's because of the way colors are coded (using three receptors in the eye, for red, green and blue, and four neural channels) doesn't explain why the *qualia* should also be chopped up into four more elementary subjective sensations. After all, once the wavelength information has been extracted (by computing ratios of activity of the three classes of cones), it could, in theory, have been represented in the brain and experienced subjectively as a continuum—just as we do with pitch. The fact that different modes of experience apply to wavelength and pitch suggests that qualia cannot be epiphenomenal; they must have an evolutionary function—such as serving as a mnemonic aid for labeling and talking about things as edible fruits (red), inedible fruits (green) vs. edible leaves (green) or sexually receptive female primate rumps (reds and blues), etc. Pitch isn't used to label things in the same way. This is, admittedly, a far-fetched argument for the distinctiveness of color qualia but it is hard to avoid being farfetched when speculating on this topic (see also Crick, 1994; Ramachandran and Hirstein, 1997; Crick and Koch, 1998). I was asked by Richard Dawkins whether bats, which "see" objects and their surface textures using echo location, might use color labels when experiencing and designating textural qualities in hearing: a not unreasonable suggestion.

Another view on introspective consciousness is that it

originally emerged in order to help simulate other people's minds—to help develop a sophisticated theory of other minds. Nick Humphrey originally proposed this at a conference I organized in Cambridge (Josephson and Ramachandran, 1979). Similar ideas have also been proposed by David Premack and Marc Hauser. At the same meeting Horace Barlow suggested an intimate link between language and consciousness.

Qualia may require a sense of self but I have difficulty accepting that it requires full-fledged language in the sense that we usually understand that term. As I noted in that same Cambridge meeting, qualia in general and colors in particular are vastly more "fine grained" than the words used to describe them.

4 It would be interesting to see how the patient would react to being poked with a needle while having an out-of-body experience. Would there be a galvanic skin response? Would the patient feel pain or simply feel detached from it all as if his *body* was experiencing pain but he was merely a spectator? The same question—about galvanic responses—could be asked of patients on Ketamine who also have out-of-body experiences.

Shai Azoulai and I recently saw a patient with a right parieto-occipital tumour who constantly experienced a visual hallucination of a twin or doppelganger always about a foot to his left and front. The twin mimed his movements in perfect synchrony. When I irrigated his left ear canal with cold water (stimulating the vestibular system) the twin was seen to "jump around" and "shrink in size" to a midget. Here is yet another reminder of how tenuous our sense of being anchored to our body really is, even though we usually take it to be one of the axioms of our existence.

5 Another ability closely linked to semantic aspects of language is symbol manipulation: the ability to juggle visual images of objects in your brain "off line."

To illustrate this I'll invent a thought experiment. (Unlike philosophers' thought experiments, this one can actually be done!) Imagine I show you three boxes of three different sizes on the floor and a desirable object dangling from a high ceiling. You will instantly stack the three boxes, with the largest one at the bottom and the smallest at the top, then climb up to retrieve the reward.

A chimp can also solve this problem, but presumably requires trial-and-error physical exploration of the boxes.

But now I modify the experiment. I add three colored luminous spots—one on each of the boxes—say red (big box), blue (intermediate box) and green (small box) and have them lying separately on the floor. I bring you into the room for the first time and expose you to the boxes long enough for you to realize which box has which spot. Then I switch the room lights off so that only the symbols of the boxes—the luminous colored dots—are visible. Finally, I bring a luminous reward into the dark room and dangle it from the ceiling. If you have a normal brain you will without hesitation put the red dot at the bottom, blue in the middle and green on the top and then climb to the top. In other words, as a human being you can create arbitrary symbols (loosely analogous to words) and then juggle them entirely in your brain, doing a virtual reality simulation to discover the solution. You could even do this if during the first phase you were shown only the red and green dotlabeled boxes and then, separately, the green and blue, and if then in the test phase I showed you the red and green dotlabeled boxes alone. (Assume that stacking even two boxes gives you better access to the reward.) I bet you could now juggle the symbols entirely in your head to establish the transitivity using conditionals—if red is bigger than blue and blue is bigger than green then red must be bigger than green—and then proceed to stack the green box on the red box in the dark to reach the reward, even though the relative size of the boxes was not currently visible.

An ape would almost certainly fail this task, which requires off-line manipulation of Sassurian (arbitrary) signs—the basis of language. But to what extent is language a *requirement* for "if/then" conditional statements done off line—especially in novel situations? What if you tried the experiment on patients with Wernicke's aphasia who have no comprehension of language? Or a Broca's aphasic who has difficulty with grammatical-function concepts such as if/then? Such experiments may go a long way in helping us explore the elusive interface between language and thought.

And what of abilities like playing chess (which requires "if/then" conditionals), doing both formal and informal algebra (John and Mary together have nine apples: John has twice as many as Mary. How many does each have?) or computer programming? Can Wernicke's and Broca's aphasics perform such tasks, assuming they were skilled chess players, mathematicians or programmers prior to the stroke? After all, formal algebra has its own "syntax" of sorts and programming, too, is a "language," but to what extent do they tap into the same brain machinery as natural language?

But isn't all this an overkill? After all, most of us can "visualize" juggling visual images without explicity using (silent) internal words like if/then—so why even invoke language? But here we have to be careful not to be deceived by introspection; it is quite possible that even what *feels* like visual symbol juggling may be tacitly using the same neural machinery as certain aspects of language, without your being aware of it.

6 But doesn't this notion of a "representation of a representation" lead to an endless regress? Wouldn't you need a third representation of the second representation as well? Not necessarily. Imagine the sentence "I know he knows that I know he stole my car." This entails a representation of his representation of my representation. But if I take it any further, I can o longer hold the representations in my head simultaneously; they start fading like an echo. (Although I can figure it out intellectually by counting.) A single metarepresentation is already a major advance and there may have simply been no selection pressure in evolution to develop this ability to unwieldy limits, given our already existing limitations of memory and attention span. Consciousness is a much more limited capacity than we usually realize.

7 What exactly does this "preparing the input into manageable chunks" for qualia, language and thought (see page 98) really entail? Here we enter the twilight zone: that magic step in evolution that transformed ape-like mentation into human consciousness and self-awareness. In *Phantoms in the Brain* I suggested that there are four functional characteristics associated with neural events that become linked to qualia as opposed to those (like blindsight) which don't. I call these the four laws of qualia: (1) indubitability/irrevocability; (2)

evocation of explicit meaning or semantic implications; (3) shortterm memory; (4) attention.

I have described the first "law" in detail in *Phantoms in the Brain*, so I will just briefly summarize the other three here. A paraplegic will have an intact knee-jerk reflex—the tapping of his tendon invariably evokes a knee jerk; but he experiences no qualia. The reason is that the sensation processed by the spinal cord is hooked up only to one output: the muscle contraction—it can't be used for anything else. A qualia-laden percept (such as seeing a yellow splotch of paint), on the other hand, has an enormous number of implications—a penumbra of associations such as "banana," "yellow teeth," "lemon," the word "yellow"—which give you the luxury of choosing which "implication" to make explicit for current needs dictated by the cingulate and other frontal structures. A choice that, in its turn, requires you to hold the information long enough in "working" memory (law 3) to allow you to deploy attention using the cingulate (law 4). That completes our examination of my "four laws of qualia."

The advantage of spelling out these criteria is that you can apply them to any system to determine whether it enjoys qualia and reflective self-awareness. (For example, would a sleepwalker have them?) It also eliminates silly questions such as "Does a Venus fly trap experience sensory 'insect quale' as it closes shut?'" (it doesn't); "Does a thermostat have temperature quale?," etc. Such questions have as little meaning to a neuroscientist as the question "Is a virus really alive?" does to a post-Watson/Crick molecular biologist. Borderline cases should be used to illuminate rather than blur distinctions.

It is fashionable for both neuroscientists and Indian mystics to assert that the self is an illusion, but if this is so then the burden of proof is on us to show *how* the illusion comes about. The clearest exposition of this problem comes from Zoltan Torey, who has made the ingenious suggestion that both the sense of self and qualia may be based on attention switching between the two cerebral hemispheres. Both he and David Darling also explore the link between language and reflective self-consciousness (Torey, 1999; Darling, 1993), just as I have done in this chapter. Torey's book, in particular, is full of dazzling new insights on many of these issues.

8 The key role played by hemispheric specialization in human consciousness has been emphasised by Marcel Kinsbourne, Jack Pettigrew, Mike Gazzaniga, Joe Bogen and Roger Sperry. Some years ago William Hirstein and I published a study showing that the non-verbal right hemisphere of a split-brain patient can lie (e.g., by non-verbally signing the wrong answer to experimenter B after receiving instructions from experimenter A to lie to B), which shows that lying doesn't require language. Bear in mind, though, that even though the right hemisphere doesn't have syntax and cannot talk, it does have some protolanguage—a rudimentary semantics and a reasonable lexicon of words that "refer" to things.

The only way finally to resolve this may be to test the left hemisphere of a split-brain patient who then has a stroke which damages the Wernicke's language area in the left hemisphere! Would his left hemisphere be capable of "off line" symbol manipulation and introspective self-consciousness? And can it lie?

We also tried testing the personality and aesthetic preferences of the two hemispheres independently using the same procedure—namely by training the right hemisphere to communicate "yes," "no" or "I don't know" non-verbally to us by picking one of three abstract shapes with the left hand. Imagine our surprise when we noticed that in patient LB the left hemisphere said it believed in God whereas the right hemisphere signaled that it was an atheist. The inter-trial consistency of this needs to be verified but at the very least it shows that the two hemispheres can simultaneously hold contradictory views on God, an observation that should send shock waves through the theological community. When a patient like this eventually dies, will one hemisphere end up in hell and the other in heaven?

9 In *Phantoms in The Brain* I suggested that when confronted with an anomaly or discrepancy, the coping styles of the two hemispheres might be fundamentally different; the left tends to smooth over them by engaging in Freudian defenses such as denials, confabulations, rationalizations, and even delusions, whereas the right hemisphere is your "reality checking" mechanism more anchored in the truth. There are hints that depressed people have, ironically, a more realistic world-view. The normal spectrum of moods and behavior in humans arises from a push-pull interaction between these antagonistic tendencies. In 1997, in the journal *Medical Hypotheses*, I also suggested that the extreme mood swings of bipolar "manic depressive" illness may reflect an actual alternation between the coping styles of the two hemispheres, raising the possibility that simple vestibular stimulatioin through cold-water caloric ear

irrigation might help restore the balance in mood by selectively activating one hemisphere or the other (Ramachandran V.S. and Pettigrew J. ; unpublished).

10 See V. S. Ramachandran, Mirror neurons and the great leap forward (1999), http://www.edge.org/3rd_culture/ramachandran/ramachandran_index.html.

11 A similar distinction between metarepresentions and representations can be made for memory. When the hippocampus is damaged on both sides of the brain, the patient develops anterograde amnesia; he can remember events that took place in the past—prior to the damage—but he is unable to form new memories. For instance, if an attending physician introduces himself to the patient, walks out to go to the restroom, and returns after five minutes, the patient has no recollection of ever having seen him. Almost a hundred years ago, the French psychologist Edouard Claparede performed an ingenious experiment. He walked in, introduced himself, and shook the patient's hand. Concealed in Claparede's palm was a pin, so the patient shouted "Ouch!" and withdrew his hand. The next day when Claparede returned, the patient had no glimmer of recognition, no inkling that he had seen Claparede before, yet when offered a handshake, he instinctively withdrew! As Endel Tulving has emphasized, such dissociations imply different memory acquisition mechanisms in the brain, only one of which depends on the hippocampus. In our terminology his brain has a rich enough representation to create vague emotional propensities such as avoidance, but it doesn't have a metarepresentation linked to his sense of self, so there is no conscious recollection of ever having seen the physician, and hence no semantic associations such as: How much money do I owe him? What's his name? Is he an M.D. or a mere Ph.D.? Can I sue him? And so on. Which raises an intriguing question: How rich is the representation that evokes reflex avoidance? (Assuming it's a memory trace that's evoked in parallel.) Can it elicit only crude emotions or behavioral propensities like avoidance, or can it also evoke— when appropriate—pity, filial love and jealousy? Will a face that *resembles* the original but is obviously perceived as differ- ent by the explicit (conscious) memory system nevertheless "fool" the tacit memory system and evoke avoidance? How sophisticated is the explicit/conscious recollection memory system in monkeys and apes, if it exists in them at all? After all, the ability to link a long string of salient "episodes" in our lives in the correct sequential order is vital for the sense of coherence and continuity of self, and it is a moot point whether the great apes have this ability.

About eight years ago, I tried waking up an amnesic patient during REM (rapid eye movement) sleep. Intriguingly, her dream reports contained specific episodic memories from the previous day even though she had no conscious recollection of these episodes when awake. This implies that the patient continues to acquire at least some new memories but without the hypocampus, she cannot index them in a consciously retrievable manner (Ramachandran 1996). Similar observations have recently been made by R. Stickgold and colleagues (2000).

The key issue regarding hypnosis is whether it is simply an elaborate form of role playing (as when you "enter " the mind of a hero, and temporarily suspend reality while watching a movie) or does it represent a genuinely different brain state (whatever that might mean)? We are exploring this by hypnotizing synesthetes into thinking they are Chinese speakers who cannot read English. Would the illegible letters and numerals then look colored nonetheless?

In more direct test Shai Azoulai and I will take advantage of the so-called size weight illusion discovered by Von Helmholtz. I give you two balls—A and B—with A being twice the size of B, but identical to B in physical weight. If you pick them up one with each hand, B will feel much heavier (almost 50 percent heavier) even though you are asked to judge absolute weight, not density. The illusion arises because you expect the larger one to be heavier, set your muscles accordingly, and are surprised when it isn't. Now imagine: I hypnotize a subject, give him two balls A and B of indentical size and weight, but tell him that he is seeing A as twice as big as B. If hypnosis is actually altering his perception and corresponding brain state, he should experience the size weight illusion, and A should feel much lighter. But if hypnosis is just extreme suggestibility he should actually say A was heavier "because its big!" Thus the two theories make opposite predictions.

Glossary

I would like to thank the Society for Neuroscience for their permission to reproduce this glossary. Some emendations have been made.

Acetylcholine A neurotransmitter both in the brain, where it may help regulate memory, and in the peripheral nervous system, where it controls the actions of skeletal and smooth muscle.

Action potential This occurs when a neuron is activated and temporarily reverses the electrical state of its interior membrane from negative to positive. This electrical charge travels along the axon to the neuron's terminal where it triggers or inhibits the release of a neurotransmitter and then disappears.

Adrenaline *see* **Epinephrine**

Affective psychosis A psychiatric disease relating to mood states. It is generally characterized by depression unrelated to events in the life of the patient, which alternates with periods of normal mood or with periods of excessive, inappropriate euphoria and mania.

Amygdala A structure in the forebrain that is an important component of the limbic system.

Anosognosia A syndrome in which a person with a paralyzed limb claims it is still functioning. (Anosognosia means denial of illness.) An explanation may involve close analysis of the different roles of the left and right hemispheres of the brain.

Antagonist A drug or other molecule that blocks receptors. Antagonists inhibit the effects of agonists.

Aphasia Disturbance in language comprehension or production, often as a result of a stroke.

Autonomic nervous system A part of the peripheral nervous system responsible for regulating the activity of internal organs. It includes the sympathetic and parasympathetic nervous systems.

Axon The fiber-like extension of a neuron by which the cell sends information to target cells.

Basal ganglia Clusters of neurons, which include the caudate nucleus, putamen, globus pallidus and substantia nigra, that are located deep in the brain and play an important role in movement. Cell death in the substantia nigra contributes to Parkinsonian signs.

Blindsight Some patients who are effectively blind because of brain damage can carry out tasks which appear to be impossible unless they can see the objects. For instance they can reach out and grasp an object, accurately describe whether a stick is vertical or horizontal, or post a letter through a narrow slot. The explanation appears to be that visual information travels along two pathways in the brain. If only one is damaged, a patient may lose the ability to see an object but still be aware of its location and orientation.

Blindspots Blindspots can be produced by a variety of factors. In fact everyone has a small blindspot in each eye caused by the area of the retina which connects to the optic nerve. These blindspots are often filled in by the brain using information based on the surrounding visual image. In some cases, patients report seeing unrelated images in their blindspots.

One reported seeing cartoon characters. This phenomenon may involve other pathways in the brain.

Brainstem The major route by which the forebrain sends information to and receives information from the spinal cord and peripheral nerves. It controls, among other things, respiration and regulation of heart rhythms.

Broca's area The brain region located in the frontal lobe of the left hemisphere that is important for the production of speech.

Capgras delusion A rare syndrome in which the patient is convinced that close relatives, usually parents, spouse, children or siblings, are imposters. It may be caused by damage to the connections between the areas of the brain dealing with face recognition and emotional response. A sufferer might recognize the faces of his loved ones but not feel the emotional reaction normally associated with the experience.

Catecholamines The neurotransmitters dopamine, epinephrine and norepinephrine that are active both in the brain and in the peripheral sympathetic nervous system. These three molecules have certain structural similarities and are part of a larger class of neurotransmitters known as monoamines.

Cerebral cortex The outermost layer of the cerebral hemispheres of the brain. It is responsible for all forms of conscious experience, including perception, emotion, thought and planning.

Cerebral hemispheres The two specialized halves of the brain. The left hemisphere is specialized for speech, writing, language and calculation; the right hemisphere is specialized for spatial abilities, face recognition in vision and some aspects of music perception and production.

Classical conditioning Learning in which a stimulus that naturally produces a specific response (unconditioned stimulus) is repeatedly paired with a neutral stimulus (conditioned stimulus). As a result, the conditioned stimulus can become able to evoke a response similar tothat of the unconditioned stimulus.

Cognition The process or processes by which an organism gains knowledge of or becomes aware of events or objects in its environment and uses that knowledge for comprehension and problem solving.

Cone A primary receptor cell for vision located in the retina. It is sensitive to color and used primarily for daytime vision.

Corpus callosum The large bundle of nerve fibers linking the left and right cerebral hemispheres.

Cotard's syndrome A disorder in which a patient asserts that he is dead, claiming to smell rotting flesh or worms crawling over his skin. It may be an exaggerated form of the Capgras delusion, in which not just one sensory area (i.e., face recognition) but all of them are cut off from the limbic system. This would lead to a complete lack of emotional contact with the world.

Dendrite A tree-like extension of the neuron cell body. Along with the cell body, it receives information from other neurons.

Dopamine A catecholamine neurotransmitter known to have multiple functions depending on where it acts. Dopamine-containing neurons in the substantia nigra of the brainstem project to the caudate nucleus and are destroyed in Parkinson's victims. Dopamine is thought to regulate emotional responses, and to play a role in schizophrenia and cocaine abuse.

Endocrine organ An organ that secretes a hormone directly into the bloodstream to regulate cellular activity of certain other organs.

Endorphins Neurotransmitters produced in the brain that generate cellular and behavioral effects similar to those generated by morphine.

Epinephrine A hormone, released by the adrenal medulla and the brain, that acts with norepinephrine to activate the sympathetic division of the autonomic nervous system. Sometimes called adrenaline.

Evoked potentials A measure of the brain's electrical activity in response to sensory stimuli. This is obtained by placing electrodes on the surface of the scalp (or, more rarely, inside the head), repeatedly administering a stimulus, and then using a computer to average the results.

Excitation A change in the electrical state of a neuron that is associated with an enhanced probability of action potentials.

Forebrain The largest division of the brain, which includes the cerebral cortex and basal ganglia. It is credited with the highest intellectual functions.

Frontal lobe One of the four divisions (the others are parietal, temporal and occipital) of each hemisphere of the cerebral cortex. It has a role in controlling movement and associating the functions of other cortical areas.

Gamma-amino butyric acid (GABA) An amino acid transmitter in the brain whose primary function is to inhibit the firing of neurons.

Glia Specialized cells that nourish and support neurons.

Hippocampus A seahorse-shaped structure located within the brain and considered an important part of the limbic system. It functions in learning, memory and emotion.

Hormones Chemical messengers secreted by endocrine glands to regulate the activity of target cells. They play a role in sexual development, calcium and bone metabolism, growth and many other activities.

Hypothalamus A complex brain structure composed of many nuclei with various functions. These include regulating the activities of internal organs, monitoring information from the autonomic nervous system and controlling the pituitary gland.

Immediate memory A phase of memory that is extremely shortlived, with information stored only for a few seconds. It also is known as short-term and working memory.

Inhibition In reference to neurons, a synaptic message that prevents the recipient cell from firing.

Ions Electrically charged atoms or molecules.

Korsakoff's syndrome A disease associated with chronic alcoholism, resulting from a deficiency of vitamin B-1. Patients sustain damage to part of the thalamus and cerebellum. Symptoms include inflammation of nerves, muttering delirium, insomnia, illusions and hallucinations and a lasting amnesia.

Limbic system A group of brain structures—including the amygdala, hippocampus, septum and basal ganglia—that work to help regulate emotion, memory and certain aspects of movement.

Long-term memory The final phase of memory in which information storage may last from hours to a lifetime.

Mania A mental disorder characterized by excessive excitement. A form of psychosis with exalted feelings, delusions of grandeur, elevated mood, psychomotor overactivity and overproduction of ideas.

Memory consolidation The physical and psychological changes that take place as the brain organizes and restructures information in order to make it a permanent part of memory.

Motor neuron A neuron that carries information from the central nervous system to the muscle.

Myelin Compact fatty material that surrounds and insulates axons of some neurons.
Neuron Nerve cell. It is specialized for the transmission of information and characterized by long fibrous projections called axons, and shorter, branch-like projections called dendrites.
Neurotransmitter A chemical released by neurons at a synapse for the purpose of relaying information via receptors.
Nociceptors In animals, nerve endings that signal the sensation of pain. In humans, they are called pain receptors.
Norepinephrine A catecholamine neurotransmitter, produced both in the brain and in the peripheral nervous system. It seems to be involved in arousal, reward and regulation of sleep and mood, and the regulation of blood pressure.
Occipital lobe One of the four subdivisions (the others are frontal, temporal and parietal) of each hemisphere of the cerebral cortex. It plays a role in vision.
Pain asymbolia People with this condition do not feel pain when, for example, stabbed in the finger with a sharp needle. Sometimes patients say they can feel the pain, but it doesn't hurt. They know they have been stabbed, but they do not experience the usual emotional reaction. The syndrome is often the result of damage to a part of the brain called the insular cortex. The stabbing sensation is received by one part of the brain, but the information is not passed on to another area, the one which normally classifies the experience as threatening and triggers—through the feeling of pain—an avoidance reaction.
Parasympathetic nervous system A branch of the autonomic nervous system concerned with the conservation of the body's energy and resources during relaxed states.

Parietal lobe One of the four subdivisions (the others are frontal, temporal and occipital) of each hemisphere of the cerebral cortex. It plays a role in sensory processes, attention and language.

Peptides Chains of amino acids that can function as neurotransmitters or hormones.

Peripheral nervous system A division of the nervous system consisting of all nerves not part of the brain or spinal cord.

Phantom limbs People who lose a limb through an accident or amputation sometimes continue to feel that it is still there. These sensations may be the result of the brain forming new connections.

Pineal gland An endocrine organ found in the brain. In some animals, it seems to serve as a light-influenced biological clock.

Pituitary gland An endocrine organ closely linked with the hypothalamus. In humans, it is composed of two lobes and secretes a number of hormones that regulate the activity of other endocrine organs in the body.

Pons A part of the hindbrain that, with other brain structures, controls respiration and regulates heart rhythms. The pons is a major route by which the forebrain sends information to and receives information from the spinal cord and peripheral nervous system.

Qualia A term for subjective sensations.

Receptor cell Specialized sensory cells designed to pick up and transmit sensory information.

Rod A sensory neuron located in the periphery of the retina. It is sensitive to light of low intensity and specialized for nighttime vision.

Serotonin A monoamine neurotransmitter believed to play many roles including, but not limited to, temperature regulation, sensory perception and the onset of sleep. Neurons using serotonin as a transmitter are found in the brain and in the gut. A number of antidepressant drugs are targeted to brain serotonin systems.

Short-term memory A phase of memory in which a limited amount of information may be held for several seconds to minutes.

Stimulus An environmental event capable of being detected by sensory receptors.

Stroke A major cause of death in the West, a stroke is an impeded blood supply to the brain. It can be caused by a blood clot forming in a blood vessel, a rupture of the blood vessel wall, an obstruction of flow caused by a clot or other material, or by pressure on a blood vessel (as by a tumor). Deprived of oxygen, which is carried by blood, nerve cells in the affected area cannot function and die. Thus, the part of the body controlled by those cells cannot function either. Stroke can result in loss of consciousness and brain function, and death.

Sympathetic nervous system A branch of the autonomic nervous system responsible for mobilizing the body's energy and resources during times of stress and arousal.

Synapse A gap between two neurons that functions as the site of information transfer from one neuron to another.

Synesthesia A condition in which a person quite literally tastes a shape or sees a color in a sound or number. This is not just a way of describing experiences as a poet might use metaphors. Synesthetes actually experience the sensations.

Temporal lobe One of the four major subdivisions (the others are frontal, parietal and occipital) of each hemisphere of the cerebral cortex. It functions in auditory perception, speech and complex visual perceptions.

Temporal lobe epilepsy A condition which may produce a heightened sense of self and which has been linked to religious or spiritual experiences. Some people may undergo striking personality, changes and may also become obsessed with abstract thoughts. One possible explanation is that repeated seizures may cause a strengthening of the connections between two areas of the brain—the temporal cortex and the amygdala. Patients have been observed to have a tendency to ascribe deep significance to everything around them, including themselves.

Thalamus A structure consisting of two egg-shaped masses of nerve tissue, each about the size of a walnut, deep within the brain. It is the key relay station for sensory information flowing into the brain, filtering out only information of particular importance from the mass of signals entering the brain.

Ventricles Of the four ventricles, comparatively large spaces filled with cerebrospinal fluid, three are located in the brain and one in the brainstem. The lateral ventricles, the two largest, are symmetrically placed above the brainstem, one in each hemisphere.

Wernicke's area A brain region responsible for the comprehension of language and the production of meaningful speech.

Bibliography

Sources

Altschuler E., Ramachandran, V. S., Pineda, J. (2000). "Mu Wave Glocking and its Use as Tool to Study Theory of Other Minds." Soc. for Neuroscience Abstracts.

Altschuler E., Wisdom, S., Stone, L., Foster, C. and Ramachandran, V. S. (1999). Rehabilitation of Hemiparesis After Stroke with a Mirror, *Lancet* 353: 2035–2036

Armel, K. C. and Ramachandran, V. S. (1999). Acquired Synesthesia in Retinitis Pigmentosa, *Neurocase* 5(4): 293–296

Baron-Cohen, S., Burt, L., Smith-Laittan, F., Harrison, J. and Bolton, P. (1996). Synaesthesia: Prevalence and Familiarity, *Perception* 25(9): 1073–1080

Berlin, B. (1994). Evidence for Pervasive Synthetic Sound Symbolism in Ethnozoological Nomenclature, in L. Hinton, J. Nichols and J.J. Ohala (eds.), *Sound symbolism*, New York: Cambridge University Press, chapter 6

Churchland, P. (1996). *Neurophilosophy*, Cambridge, MA: MIT Press
(2002). *Brain Wise: Studies in Neurophilosophy*, Cambridge, MA: MIT Press

Clarke, S., Regali, L., Janser, R. C., Assal, G. and De Tribolet, N. (1996). Phantom Face, *Neuroreport* 7: 2853–2857

Crick, F. (1994). *The Astonishing Hypothesis: The Scientific Search for the Soul*, New York: Scribner

Crick, F. and Koch, C. (1998). Consciousness and Neuroscience, *Cerebral Cortex* 8(2): 97–107

Darling, D. (1993). *Equations of Eternity*, New York: MJF Books

Deacon, T. (1997). *The symbolic Species*, Harmondsworth: Penguin

Domino, G. (1989). Synesthesia and Creativity in Fine Arts Students: an Empirical Look, *Creativity Research Journal* 2(1–2): 17–29

Ellis, H., Young, A. W., Quale, A. H. and De Pauw, K. W. (1997). Reduced Autonomic Responses to faces in Capgras Syndrome, *Proceedings of the Royal Society of London* B 264: 1085–1092

Franz, E. and Ramachandran, V. S. (1998). Bimanual Coupling in Amputees with Phantom Limbs, *Nature Neuroscience* 1: 443–444

Frith, C. and Dolan, R. (1997). Abnormal Beliefs, Delusions and Memory. Conference presentation, Harvard conference on memory and belief

Galton, F. ([1880]1997). Colour Associations, in S. Baron-Cohen and J. E. Harrison (eds.), *Synaesthesia: Classic and Contemporary Readings*, Oxford: Blackwell, pp. 43–48

Greenfield, S. (2002). *Private Life of the Brain*, Harmondsworth: Penguin

Harris, A. J. (1999). Cortical Origin of Pathological Pain, *Lancet* 354: 1464–1466

Hirstein, W. and Ramachandran, W. S. (1997). Capgras Syndrome, *Proceedings of the Royal Society of London* B 264: 437–444

Humphrey, N. (1983). *Consciousness Regained*, Oxford: Oxford University Press

Hurley, S. and Noe, A. (2003). Neural Plasticity and Consciousness, *Biology and Philosophy* 18: 131–168

Josephson, B. and Ramachandran, V. S. (1979). *Consciousness and the Physical World*, Oxford: Pergamon Press

La Cerra, P. and Bingham, R. (2002). *The Origin of Minds: Evolution, Uniqueness, and the New Science of the Self*, New York: Harmony Books

Lakoff, G. and Johnson, M. (1999). *Philosophy in the Flesh: The Embodied Mind and Its Challenge to Western Thought*, New York: Basic Books

Lueck, C. J., Zeki, S., Friston, K. J., Deiber, M. P., Cope, P., Cunningham, V. J., Lammertsma, A. A., Kennard, C. and Frackowiak, R. S. (1989). The Colour Centre in the Cerebral Cortex of Man, *Nature* 340: 386–389

McCabe, C. S., Haigh, R. C., Ring, E. F., Halligan, P., Wall, P. D. and Blake, D. R. (2003). A Controlled Pilot Study of the Utility of Mirror Visual Feedback in the Treatment of Complex Regional Pain Syndrome (type 1), *Rehematology* 42: 97–101

Melzack, R. (1992). Phantom Limbs, *Scientific American* 266: 120–126

Merikle, P., Dixon, M. J. and Smilek, D. (2002). The Role of Synaesthetic Photisms on Perception, Conception and Memory. Speech Delivered at the 12th Annual Meeting of the Cognitive Neuroscience Society, San Francisco, CA, 14–16 April

Merzenich, M. and Kaas, J. (1980). Reorganization of Mammalian Somatosensory Cortex Following Peripheral Nerve Injury, *Trends in Neuroscience* 5: 434–436

Miller, S. and Pettigrew, J. D. (2000). Interhemispheric Switching Mediates Binocular Rivalry, *Current Biology* 10: 383–392

Nielsen, T. L. (1963). Volition: A New Experimental Approach, *Scandinavian Journal of Psychology* 4: 215–230

Nunn, J. A., Gregory, L. J., Brammer, M., Williams, S. C. R., Parslow, D. M., Morgan, M. J., Morris, R. G., Bullmore, E. T., Baron-Cohen, S. and Gray, J. A. (2002). Functional Magnetic Resonance Imaging of Synesthesia: Activation of V4/V8 by Spoken Words, *Nature Neuroscience* 5(4): 371–375

Pons, T. P., Garraghty, P. E., Ommaya, A. K., Kaas, J., Taub, E. and Mischkin, M. (1991). Massive Cortical Reorganization After Sensory Deafferentation in Adult Macaques, *Science* 252: 1857–1860

Ramachandran, V. S. (1995). Anosognosia, *Consciousness and Cognition* 1: 22–46

(2001). Sharpening Up "the Science of Art," *Journal of Consciousness Studies* 8(1): 9–29

(1996). Illusion of body image, in *The Mind Brain Continuum*, ed R. Llinas and P. F. Churchland, Boston: MIT Press.

Ramachandran, V. S., Altschuler, E. and Hillyer, S. (1997). Mirror agnosia, *Proceedings of the Royal Society of London* B 264: 645–647

Ramachandran, V. S. and Blakeslee, S. (1998). *Phantoms in the Brain*, New York: William Morrow

Ramachandran, V. S. and Hirstein, W. (1997). Three Laws of Qualia: What Neurology Tells Us about the Biological Functions of Consciousness, *Journal of Consciousness Studies* 4(5–6): 429–457

(1998). The Perception of Phantom Limbs; the D. O. Hebb Lecture, *Brain* 121: 1603–1630

(1999). The Science of Art: A Neurological Theory of Aesthetic Experience, *Journal of Consciousness Studies* 6(6–7): 15–51

Ramachandran, V. S. and Hubbard, E. M. (2001a). Psychophysical Investigations into the Neural Basis of Synaesthesia, *Proceedings of the Royal Society of London* B 268: 979–983

(2001b). Synaesthesia—a Window into Perception, Thought and Language, *Journal of Consciousness Studies* 8(12): 3–34

(2002). Synesthetic Colors Support Symmetry Perception, Apparent Motion, and Aambiguous Crowding. Speech Delivered at the 43rd Annual Meeting of the Psychonomics Society, 21–24 November

(2003). Hearing Colors and Tasting Shapes, *Scientific American*, May: 52–59

Ramachandran, V. S. and Rogers-Ramachandran, D. (1996). Denial of Disabilities in Anosognosia, *Nature* 377: 489–490

Ramachandran, V. S., Rogers-Ramachandran, D. and Stewart, M. (1992). Perceptual Correlates of Massive Cortical Reorganization, *Science* 258: 1159–1160

Sathian, K., Greenspan, A. I. and Wolf, S. L. (2000). Doing It with Mirrors: A Case Study of a Novel Approach to Rehabilitation, *Neurorehabilitation and Neural Repair* 14: 73–76

Shödinger, Erwin (1992). Mind and Matter, in *What Is Life?*, New York: Cambridge University Press

Stevens, J. and Stoykov, M. E. (2003). Using Motor Imagery in the Rehabilitation of Hemiparesis, *Archives of Physical and Medical Rehabilitation* 84: 1090–1092

Strickgold, R., A. Malia, D. Maguire. D. Roddenberry, M. O'Connor (2000). "Hypnagogic Images in Normals and Amnesiacs." *Science* 290: 350-353.

Stoerig, P. and Cowey, A. (1989). Wavelength Sensitivity in Blindsight, *Nature* 342: 916–918

Torey, Z. (1999). *The Crucible of Consciousness*, Oxford: Oxford University Press

Treisman, A. M. and Gelade, G. (1980). A Feature-Integration Theory of Attention, *Cognitive Psychology* 12(1): 97–136

Turton, A. J. and Butler, S. R. (2001). Referred Sensations Following Stroke, *Neurocase* 7(5): 397–405

Wegner. D. (2002). *The Illusion of Conscious Will*, Cambridge, MA: MIT Press

Weiskrantz, L. (1986). *Blindsight*, Oxford: Oxford University Press

Whiten, A. (1998). Imitation of Sequential Structure of Actions in Chimpanzees, *Journal of Comparative Psychology* 112: 270–281

Young, A. W., Ellis, H. D., Quayle, A. H. and De Pauw, K. W. (1993). Face Processing Impairments and the Capgras Delusion, *British Journal of Psychiatry* 162: 695–698

Zeki, S. and Marini, L. (1998). Three Cortical Stages of Colour Processing in the Human Brain, *Brain* 121(9): 1669–1685

General reading

Baddeley, A. D. (1986). *Working Memory*, Oxford: Churchill Livingstone

Barlow, H. B. (1987). *The Biological Role of Consciousness in Mindwaves 361–381*, Oxford: Basil Blackwell

Baron-Cohen, S. (1995). *Mindblindness*, Cambridge, MA: MIT Press

Bickerton, D. (1994). *Language and Human Behaviour*, Seattle: University of Washington Press

Blackmore, Susan. (2003). *Consciousness: An Introduction*, New York: Oxford University Press

Blakemore, C. (1997). *Mechanics of Mind*, Cambridge: Cambridge University Press

Carter, R. (2003). *Exploring Consciousness*, Berkeley: University of California Press

Chalmers, D. (1996). *The Conscious Mind*, New York: Oxford University Press

Corballis, M. C. (2002). *From Hand to Mouth: The Origins of Language*, Princeton: Princeton University Press

Crick, F. (1993). *The Astonishing Hypothesis*, New York: Scribner

Cytowick, R. E. (2002). *Synaesthesia: A Union of the Senses*, 2nd Edition (originally published 1989), New York: Springer-Verlag

Damasio, A. (1994). *Descartes' Error*, New York: G. P. Putnam

Dehaene, S. (1997). *The Number Sense: How the Mind Creates Mathematics*, New York: Oxford University Press

Dennett, D. C. (1991). *Consciousness Explained*, New York: Little, Brown, and Co.

Edelman, G. M. (1989). *The Remembered Present: A Biological Theory of Consciousness*, New York: Basic Books

Ehrlich, P. (2000). *Human Natures*, Harmondsworth: Penguin Books

Gazzaniga, M. (1992). *Nature's Mind*, New York: Basic Books

Glynn, I. (1999). *An Anatomy of Thought*, London: Weidenfeld and Nicolson

Greenfield, S. (2000). *The Human Brain: A Guided Tour*, London: Weidenfeld and Nicolson

Gregory, R. L. (1966). *Eye and Brain*, London: Weidenfeld and Nicolson

Hubel, D. (1988). *Eye, Brain and Vision*, New York: W. H. Freeman

Humphrey, N. (1992). *A History of the Mind*, New York: Simon and Schuster

Kandel er Schwartz, J. and Jessel, T. M. (1991). *Principles of Neural Science*, New York: Elsevier

Kinsbourne, M. (1982). Hemispheric specialization, *American Psychologist* 37: 222–231

Milner, D. and Goodale, M. (1995). *The Visual Brain in Action*, New York: Oxford University Press

Mithen, Steven. (1999). *The Prehistory of the Mind*, London: Thames & Hudson

Pinker, S. (1997). *How the Mind Works*, New York: W. W. Norton

Posner, M. and Raichle, M. (1997). *Images of Mind*, New York: W. H. Freeman

Premack, D. and Premack, A. (2003). *Original Intelligence*, New York: McGraw-Hill

Quartz, S. and Sejnowski, T. (2002). *Liars, Lovers and Heroes*, New York: William Morrow

Robertson, I. (2000). *Mind Sculpture*, New York: Bantam

Sacks, O. (1985). *The Man Who Mistook His Wife for a Hat*, New York: HarperCollins

(1995). *An Anthropologist on Mars*, New York: Alfred Knopf

Schacter, D. L. (1996). *Searching for Memory*, New York: Basic Books

Wolpert, L. (2001). *Malignant Sadness: The Anatomy of Depression*, Faber and Faber

Zeki, S. (1993). *A Vision of the Brain*, Oxford: Oxford University Press

Acknowledgments

First, I must thank my parents, who always nurtured my curiosity and interest in science. My father bought me a Zeiss microscope when I was eleven and my mother helped me set up a chemistry lab under the staircase in our house in Bangkok, Thailand. Many of my teachers at the Bangkok British (Patana) school, especially Mrs. Vanit and Mrs. Panachura, gave me chemicals to take home and "experiment" with.

My brother V. S. Ravi played an important role in my early upbringing: he would often recite Shakespeare and the *Rubaiyat* to me. Poetry and literature have a great deal more in common with science than people realize; they both involve unusual juxtapositions of ideas and a certain "romantic" view of the world.

My thanks to Semmangudi Sreenivasa Iyer, whose divine music was always a tremendous catalyst to all my endeavors.

Gratitude to Jayakrishna, Chandramani and Diane for being a constant source of delight and inspiration; to the BBC Reith lecture staff Gwyneth Williams and Charles Siegler for the excellent job they did in editing the lectures and to Sue Lawley for hosting the events; to the staff of Profile Books, Andrew Franklin and Penny Daniel, who helped transform the lectures into a readable book for the United Kingdom. And for the

North American edition, I thank Stephen Morrow and Pi Press for their good creative work.

Science flourishes best in an atmosphere of complete freedom and financial independence. No wonder it reached its zenith during times of great prosperity and patronage of learning—in ancient Greece, where the science of logic and geometry first emerged; in the golden age of the Guptas in India (around the fifth century AD), when the number system, trigonometry and much of algebra as we now know it were born; and during the Victorian era—the era of gentleman scientists like Humphry Davy, Darwin and Cavendish. The closest thing to this that we now have in the United States is the tenure system and federally funded grants, and I am especially grateful to the NIH for having provided uninterrupted support for my research over the years. (But, as many of my students have learned, the system isn't perfect, often unwittingly rewarding the conformist and punishing the visionary. As Sherlock Holmes told Watson, "Mediocrity knows nothing higher than itself; it requires talent to recognize genius.")

My career as a medical student was strongly influenced by six prominent physicians: K.V. Thiruvengadam, P. Krishnan Kutti, M. K. Mani, Sharada Menon, Krishnamurti Sreenivasan and Rama Mani. Later, when I went up to Trinity College, Cambridge, I found myself in a very intellectually stimulating environment. I remember the many conversations with other research students and colleagues: Sudarshan Iyengar, Ranjit Kumar Nair, Mushirul Hasan, Hemal Jayasurya, Hari Vasdudevan, Arfaei Hessam, and Vidya and Prakash Virkar.

Among the teachers and colleagues who influenced me most

I should mention Jack Pettigrew, Richard Gregory, John Allman, Oliver Sacks, Horace Barlow, Sir Alan Gilchrist, Dave Peterzell, Edie Munk, P. C. Anand Kumar, Sheshagiri Rao, T. R. Vidyasagar, V. Madhusudhan Rao, Vivian Barron, Oliver Braddick, Fergus Campbell, C. C. D. Shute, Colin Blakemore, David Whitteridge, Donald McKay, Don McLeod, David Presti, Alladi Venkatesh, Carrie Armell, Ed Hubbard, Eric Altschuler, Ingrid Olson, Pavithra Krishnan, David Hubel, Ken Nakayama, Marge Livingstone, Nick Humphrey, Brian Josephson, Pat Cavanagh, Bill Hurburt and Bill Hirstein. I have also maintained strong links with Oxford over the years: Ed Rolls, Anne Treisman, Larry Weiskrantz, John Marshall and Peter Halligan. I am grateful to All Souls College for electing me to a fellowship in 1998, an affiliation that is unique in that it carries no formal responsibilities of any kind (indeed, too much hard work is frowned upon). It gave me the leisure to think and write about the neurology of aesthetics—the topic of my third Reith lecture. My interest in art was also fueled by Julia Kindy, a brilliant art historian at UCSD. Her inspiring courses on Rodin and Picasso got me thinking about the science of art.

Thanks to the Athenaeum Club, which provided excellent library facilities and a safe haven whenever I wanted to escape the hustle and bustle of the city during my visits to London; and to Esmeralda Jahan, the eternal muse to all aspiring scientists and artists.

I also had the good fortune of having many uncles and cousins who are distinguished scientists and engineers. I thank my uncle, Alladi Ramakrishnan, an eminent physicist who encouraged my early interest in science; when I was nineteen years old he had his secretary Ganapathy type my manuscript on stereopsis for *Nature* and, to my amazement (and his!), it was accepted and published without revision. The physicist P. Hariharan—one of the inventors of white light holography—had a major influence on my early intellectual development. I have also enjoyed many stimulating conversations with Alladi Prabhakar, an outstanding scholar in the field of telecommunications who has inspired many generations of students; Krishnaswami Alladi, a wizard in number theory; Ishwar (Isha) Hariharan, whose rapid rise to preeminence in the highly competitive field of molecular biology I have watched with avuncular pride (and who has now joined the UC system, I'm happy to say); and Kumpati Narendra, the first of the Alladi cousins to defy Brahminical prohibitions against crossing the oceans to the United States—where he became a leading expert in the field of control theory and systems engineering.

Other friends, relatives and colleagues: Shai Azoulai, Liz Bates, Roger Bingham, Jeremy Brockes, Steve Cobb, Nikki De Saint Phalle, Gerry Edelman, Rosetta Ellis, Jeff Ellman, K. Ganapathy, Lakshmi Hariharan, Bela Julesz, Kristof Koch, Dorothy Kleffner, P. C. Anand Kumar, S. Lakshmanan, Steve Link, Kumpati Narendra, Malini Parathasarathy, Hal Pashler, Dan Plummer, R. K. Raghavan, K. Ramesh, Ravi (editor of *The Hindu*), Bill Rosar, Krish Sathian, Spencer Seetaram, Terry Sejnowski, Chetan Shah, Gordon Shaw, Lindsey Shenk, Alan Snyder, A.V. Sreenivasan, Subramanian Sriram, K. Sriram, Claude Valenti, Ajit Varki, Alladi Venkatesh, Nairobi Venkatraman and Ben Williams, many of whom hosted my visits to Madras.

Special thanks to Francis Crick, who at eighty-six continued to be more ebullient and passionate about science than most of my junior colleagues. Also to Stuart Anstis, a distinguished vision scientist who has been my friend and collaborator for over two decades. And to Pat and Paul Churchland, Leah Levi and Lance Stone, my colleagues here at UCSD. It also helps to have enlightened administrators and chairpersons such as Paul Drake, Jim Kulik, John Wixted, Jeff Ellman, Robert Dynes and Marsha Chandler. Financial support for research has come from Richard Geckler and Charlie Robins, who have, over the years, taken a keen interest in the work being done at our center.

Index

O

P

Q

R

S

T

About the Author

V.S. Ramachandran M.D., Ph.D., is director of the Center for Brain and Cognition and professor of psychology and neuroscience at the University of California, San Diego, and adjunct professor of biology at the Salk Institute. He trained as a physician and later obtained a Ph.D. from Trinity College at the University of Cambridge. He has received many honors and awards, including a fellowship from All Soul's College, Oxford, an honorary doctorate from Connecticut College, the Ariens Kappers Gold Medal from the Royal Nederlands Academy of Sciences, for landmark contributions in neuroscience, a gold medal from the Australian National University, a fellowship from the Society for Experimental Psychology (USA), the presidential lecture award from the American Academy of Neurology and the Ramon Cajal award from the International Neuropsychiatry Society. He gave the Decade of the Brain lecture at the twenty-fifth annual (Silver Jubilee) meeting of the Society for Neuroscience (1995), the inaugural keynote lecture at the Decade of the Brain conference held by NIMH at the Library of Congress, the Dorcas Cumming Lecture at Cold Spring Harbor, the Raymond Adams Lecture at Massachusetts General Hospital, Harvard, and the Jonas Salk memorial lecture at the Salk Institute. Dr. Ramachandran has published over 120 papers in scientific journals (including four invited review articles in *Scientific American*), and is author of the critically acclaimed book *Phantoms in the Brain*, which has been translated into eight languages and formed the basis for a two-part series on Channel 4 TV in the UK and a one-hour PBS special in the USA. *Newsweek* has named him a member of "the century club": one of the hundred most prominent people to watch in the twenty-first century.